Laser plasma accelerators for charged particles

Dissertation zur Erlangung des akademischen Grades

Doctor rerum naturalium (Dr. rer. nat.)

FRIEDRICH-SCHILLER-UNIVERSITÄT JENA
PHYSIKALISCH-ASTRONOMISCHE FAKULTÄT

vorgelegt dem Rat der Physikalisch-Astronomischen Fakultät
der Friedrich-Schiller-Universität Jena

von Dipl.–Phys. Kay-Uwe Amthor,

geboren am 10. April 1977 in Cottbus

Bibliographische Information der Deutschen Bibliothek: Die Deutsche Bibliothek verzeichnet diese Publikation in der Deutschen Nationalbibliographie; detaillierte bibliographische Daten sind im Internet über `http://dnb.ddb.de` abrufbar.

©2006 Kay-Uwe Amthor
Herstellung und Verlag: Books on Demand GmbH, Norderstedt
Satz und Layout durch den Autor mit LaTeX2e und Koma-Script.
ISBN 978-3-8334-7087-5

Contents

1 Introduction   **1**

2 Theory of laser plasma interaction   **4**
   2.1 Propagation of electromagnetic waves in plasmas 6
      2.1.1 Linear propagation 7
      2.1.2 Nonlinear propagation, self-focusing and channel formation   8
   2.2 Laser acceleration of electrons 12
      2.2.1 Laser wakefield acceleration 12
      2.2.2 Direct laser acceleration 14
      2.2.3 Bubble acceleration 14
   2.3 Laser acceleration of protons and ions 15
      2.3.1 Laser plasma interaction with solid targets 16
      2.3.2 Target Normal Sheath Acceleration 17

3 Electron Acceleration   **23**
   3.1 JETI – The Jena Titanium:Sapphire Laser 23
   3.2 Experimental setup 25
   3.3 Channel observation 30
   3.4 Simulation of laser plasma interaction 34
   3.5 Electron emission 36
      3.5.1 Quasi monoenergetic electron bunches 37
      3.5.2 Electron spectra 39
   3.6 Tracking plasma bubbles 48
      3.6.1 Shadow images 49
      3.6.2 Ring Structures 51
      3.6.3 Raytracing for shadowimages 53

4 Proton Acceleration   **58**
   4.1 Experimental setup 58
   4.2 Protons from plain targets 61
   4.3 Protons from coated and microstructured targets 63
   4.4 Simulation and scalability 69

5 Conclusion and Applications   **74**
    5.1   A future option for particle acceleration   74
    5.2   Applications for laser accelerated electrons   76
    5.3   Applications for laser accelerated protons/ions   82

Bibliography   **87**

§ 1   Introduction

The interaction of high intensity laser pulses with matter yields a copious amount of exciting and interesting physics and applications. Laser generated plasmas using high intensity femtosecond pulses have proven to be a versatile source of particles, e.g. electrons [1, 2], protons and heavy ions [3–6], as well as short pulses of electromagnetic radiation in the energy ranges of extreme UV, x-rays and γ-radiation [7–9], which opened up a broad variety of new applications, for example laser induced nuclear reactions and isotope production [10–13] and time-resolved x-ray diffraction [14]. From high-intensity laser plasmas also neutrons have been generated by fusion of deuterium [15, 16]. A summary of processes involved is given in Fig. 1.1.

The initial mechanism for all the processes conceivable in laser-plasma interaction is the generation of a plasma by a high intensity laser pulse and the acceleration of electrons by the laser inside this plasma. From the hot plasma high energy line emission in the EUV range may be observed. The plasma may reach temperatures, sufficient for fusion of deuterium nuclei for deuterated targets leading to neutron emission. The fast electrons penetrating a solid target may in turn cause x-ray line emission and bremsstrahlung, which reaches high photon-energies suitable to induce photo-nuclear reactions.

The nature of the experiment is determined strongly by the use of the target. The laser plasma interaction in a gaseous target can nowadays supply electron beams with energies of several hundred MeV, gained only within short distances of some hundred µm. At the *Institut für Optik und Quantenelektronik* a laser plasma accelerator was implemented in order to investigate the underlying mechanisms for electron acceleration in the given parameter range by using the JETI laser system to provide the high intensity pulses, and a pulsed helium gas jet acting as target. The electrons generated with this laser plasma accelerator have been used in experiments to induce photo-nuclear reactions [17], build the "photon collider" [18], which demonstrated the first autocorrelation at relativistic intensities [19], as well as the first generation of Thomson backscattered x-rays in an all optical setup [20].

Recent experiments have shown that the interaction of a high power sub–60 fs pulse with underdense gas jets leads to quasi monoenergetic electron bunches with drastically increased efficiency and controllability of electron acceleration [21–23]. The underlying acceleration mechanism is called bubble acceleration, blow-out

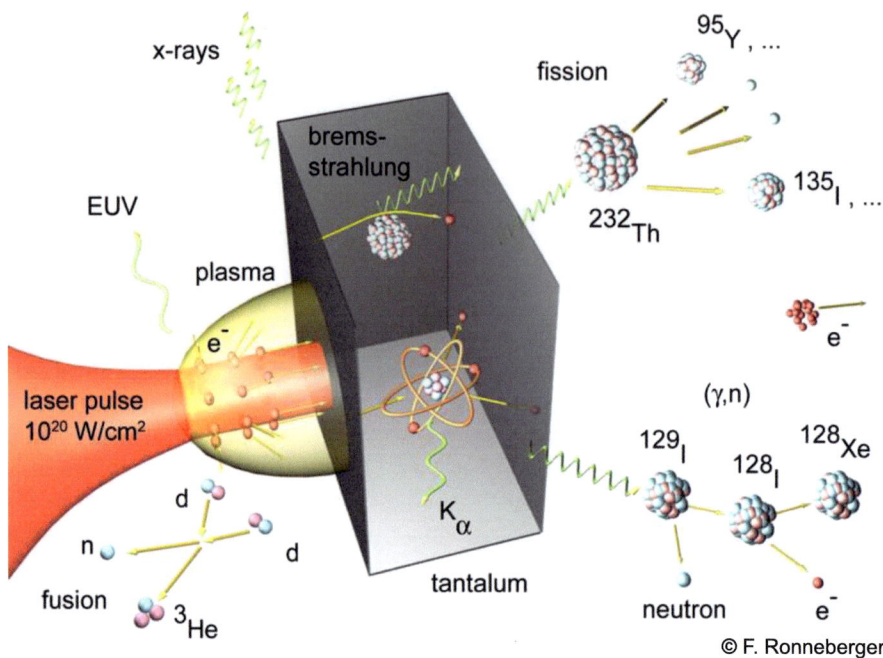

Figure 1.1: Laser plasma interaction and subsequent processes. A high intensity laser pulse generates a plasma in front of a solid target. Depending on the target material, EUV and fusion neutrons may be emitted from the hot plasma. Electrons are accelerated into the target, where x-ray line emission and bremsstrahlung will be generated, which in turn may be used to trigger photo-nuclear reactions. (Artwork courtesy of F. Ronneberger)

regime or broken wave regime and was first predicted by Pukhov and Meyer-ter-Vehn [24] via numerical simulations and was well analyzed both theoretically and numerically in recent years [25]. With the work presented in this thesis, it was demonstrated that the broken wave regime may be reached by self-modulated laser wakefield acceleration in a relativistic channel, using a different parameter range – higher plasma densities, longer laser pulses [26]. Furthermore, the temporal evolution of the electron acceleration inside the relativistic channel was investigated and the bunching of the electrons was observed directly [27].

Electron acceleration also takes place during the interaction of a high intensity laser pulse with a solid target, e. g. foils several micrometers thin. Here, the acceleration relies upon a combination of laser wakefield acceleration and/or direct laser acceleration in the plasma that forms at the front side of the thin foil. The electrons are prominently accelerated through the foil, leading to strong electric fields between the electrons and the target protons/ions. In laser foil interactions the gradients are significantly higher than for gas jets. Therefore, foils have been identified to serve as a sophisticated source of high energy protons and ions [3, 4, 6, 28].

The laser plasma accelerator at our institute was modified to work effectively for proton/ion-acceleration. The pulsed gas jet was replaced by thin metal foils, and influence of the target design on the proton acceleration was investigated. It turned out, that by using microstructured targets, the acceleration can be tailored to generate protons with narrow-band spectra [29]. The surge of interest in adapting the spectra form laser accelerated ions to be suitable for further applications is indicated by a number of other papers published early this year [30, 31].

Given the fast progress in laser development and the research on laser plasma accelerators, their characteristics may be assumed to improve over the next years to such a level, that especially the laser plasma accelerators based on table top systems with high repetition rates will become of more importance as a new generation of particle injectors for accelerators, light sources of short duration and high brilliance, and source of proton and ion beams, that may be even suitable for medical radiation therapy [12, 32, 33].

The separate implementations of the laser plasma accelerator for electrons and ions already indicates the overall structure of this thesis. The first part will concentrate on the experiments for laser acceleration of electrons from a gas jet, where the generation of quasi-monoenergetic electrons was achieved and the first direct observation of the bubble associated to the broken wave regime was demonstrated. The second part is dedicated to our experimental findings on proton and ion acceleration in laser plasmas, leading to the generation of quasi-monoenergetic protons. Before going into detail of the experiments in Chap. 3 and Chap. 4, some theoretical background for the laser plasma acceleration of electrons and ions to the extend necessary for analysis and discussion of the experimental findings, will be introduced, in the next chapter. Finally, some selected applications for laser based accelerators presented here, are explicated in the last chapter.

§ 2 Theory of laser plasma interaction

Talking about high intensities means that laser pulses at intensities in the range of $I_L = 10^{18} \ldots 10^{20}$ W/cm^2 have to be considered. The electric field of such laser pulses can reach values of the order of $E_L \sim 3 \times 10^{10} \ldots 3 \times 10^{11}$ V/cm, as compared to the atomic electric field of $E_{at} \sim 5 \times 10^9$ V/cm, which is the electric field experienced by an electron on the first orbit in the Bohr model. This means that matter – solid, liquid or gaseous – is already fully or partially ionized by the leading edge of the laser pulse and the high intensity part of the pulse is interacting with a plasma. Consequently, the electrons are subject to the highest electromagnetic fields that can be generated on earth.

A free electron in an alternating electric field $\vec{E} = E_0 \hat{e}_x \cos(\omega t - kz)$ of frequency ω oscillates with a classical velocity amplitude, also called quiver velocity, of

$$v_{\mathrm{osc,class}} = \frac{eE}{m_0 \omega} \tag{2.1}$$

If the classical quiver velocity becomes of the order of the light velocity c or higher, one has reached the relativistic interaction regime. The dimensionless amplitude a_0 serves as a parameter to determine the nature of the interaction. It is calculated as

$$a_0 = \frac{v_{\mathrm{osc,class}}}{c} = \frac{eE}{m_0 \omega c} . \tag{2.2}$$

The parameter can also be expressed by laser intensity I_L and wavelength λ directly in practical units

$$a_0 = \sqrt{\frac{I_L \times (\lambda[\mu m])^2}{1.37 \times 10^{18} \text{ W/cm}^2}} . \tag{2.3}$$

This means, for laser pulses at a wavelength of $\lambda = 0.795$ µm with intensities $I_L = 10^{18} \ldots 10^{20}$ W/cm^2 the dimensionless amplitude reaches values of $a_0 \sim 0.7 \ldots 7$, indicating laser interaction in a range between the weakly and strongly relativistic regime.

Of course, the high electron velocities in this regime will have the consequence that the electromagnetic force exerted on an electron is not longer dominated by the electric field alone. Now also the magnetic field in the Lorentz force law has to be considered

$$\vec{F} = \frac{\mathrm{d}}{\mathrm{d}t}(\gamma m \vec{v}) = -e \left(\vec{E} + \vec{v} \times \vec{B} \right) , \qquad (2.4)$$

where $\gamma = (1 - v^2/c^2)^{-1/2}$ is the relativistic factor. In an electromagnetic wave propagating in z-direction the $\vec{v} \times \vec{B}$ – term of the Lorentz force will lead to acceleration of the electron in the same direction. This additional relativistic effect leads to nonlinear Thomson-scattering of laser light from electrons and serves as a very important tool for the diagnostic of relativistic plasma channels [17, 26, 27].

Nonlinear Thomson scattering

The motion of an electron in a plane electromagnetic wave propagating in z-direction $\vec{E} = E_0 \hat{e}_x \cos(\omega t - kz)$ at relativistic intensities may be obtained by solving Eq. (2.4). Analytical solutions are demonstrated in great detail in the literature, e. g. [18, 34–36]. Therefore, just the results for the electron trajectories will be stated here. For an electron initially at rest the relativistic motion in such a field may be described by

$$\begin{aligned} k\,x &= a_0(1 - \cos\phi) , \\ k\,z &= \frac{a_0{}^2}{4}(\phi - \frac{1}{2}\sin 2\phi) , \end{aligned} \qquad (2.5)$$

where $\phi = \omega t - kz$ is the phase of the electromagnetic field. The contribution from the B-field now yields a drift in z-direction with an average velocity of

$$\frac{v_{\mathrm{D}}}{c} = \frac{a_0{}^2}{4 + a_0{}^2} . \qquad (2.6)$$

In the electron's rest frame the motion results in a trajectory shaped like a "Figure-8" (Fig. 2.1). The oscillation in z-direction in Eq. (2.5) occurs at twice the laser frequency. A rigorous treatment for the spectrum from nonlinear Thomson-scattering emitted in x-direction is given in [18], and concludes that for high laser field strength also higher harmonics contribute to the emission around 2ω.

Using Liénard-Wiechert potentials [37] the emission characteristics may be calculated from the electron trajectories. This is known as nonlinear Thomson scattering [38–41] and has been exploited to full extend in performing the first auto-correlation measurements at relativistic intensities at our laboratory as presented in [18, 19]. For the purpose of the results presented in this thesis, the intuitive picture of electrons, subject to relativistic laser pulses, emitting light around the

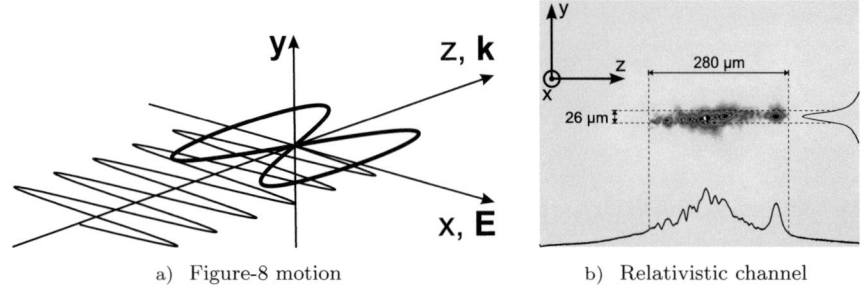

a) Figure-8 motion   b) Relativistic channel

Figure 2.1: Figure-8 motion and observation of the emission from nonlinear Thomson-scattering. The electromagnetic wave propagating in z-direction is polarized in x-direction and the electron will oscillate in the x-z-plane. The second harmonic emission may be observed along the x-axis.

second harmonic frequency is sufficient. This emission will serve as a criterion for the observation and measurement of relativistic plasma channels (Sec. 3.3).

2.1. Propagation of electromagnetic waves in plasmas

For the investigation of laser plasma interaction, the collective behavior of $n_e \sim 10^{19} \ldots 10^{23}$ electrons/cm^3 instead of a single electron has to be considered. A plasma does not consist of the electrons alone but is usually described as a composition of an electron- and ion-fluid, with electron and ion density $n_e = Z n_i$, where Z is the charge state of the ions. The Coulomb potential of a single ion with charge Ze is shielded by surrounding electrons and therefore modified to [42, 43]

$$\Phi_{\mathrm{ion}}(r) = \frac{1}{4\pi\varepsilon_0} \frac{Ze}{r} \exp\left[-\frac{r}{\lambda_D}\right] \tag{2.7}$$

with λ_D the Debye length

$$\lambda_D = \sqrt{\frac{\varepsilon_0 T_e}{n_e e^2}} \, , \tag{2.8}$$

where T_e is the plasma temperature. Throughout this thesis the plasma temperature T_e will be used. It is actually rather an energy equivalent than a temperature, because Boltzmann's constant is set to unity here, $k_B = 1$. The modification of the ion Coulomb potential by its surrounding electrons is known as Debye shielding, meaning that on scales larger than the Debye length the plasma can be considered quasi neutral.

Small-scale density deviations in a plasma lead to electrostatic forces due to space charge separation. Therefore, electron and ion sheets will oscillate at frequencies determined by their densities:

$$\omega_{\text{p,e}} = \sqrt{\frac{n_e e^2}{\varepsilon_0 m_e}}\,, \qquad \omega_{\text{p,i}} = \sqrt{\frac{n_i (Ze)^2}{\varepsilon_0 m_i}} = \omega_{\text{p,e}} \sqrt{\frac{Z m_e}{m_i}}\,, \qquad (2.9)$$

respectively.

2.1.1. Linear propagation

Owing to the much higher mass of the ions, the electron plasma waves oscillate much faster than the ion plasma waves $\omega_{\text{p,e}} \gg \omega_{\text{p,i}}$. Thus, only the electrons have to be taken into account for the response of a plasma to light and the plasma will be represented by the plasma frequency $\omega_{\text{p}} = \omega_{\text{p,e}}$. The linear dispersion relation for an electromagnetic wave propagating through a plasma reads

$$\omega^2 = \omega_p{}^2 + c^2 k^2\,, \qquad (2.10)$$

where ω and k are the frequency and wavenumber of the electromagnetic wave, respectively. Therefore, the refractive index of the plasma is given by

$$\eta = ck/\omega = \sqrt{1 - \omega_p{}^2/\omega^2} = \sqrt{1 - n_e/n_c} \qquad (2.11)$$

Here the critical plasma density n_c for an electromagnetic wave has been introduced

$$n_c = \frac{\varepsilon_0 m_e \omega^2}{e^2} = \frac{1.1 \times 10^{21}}{(\lambda/\mu\text{m})^2}\ \text{cm}^{-3}\,, \qquad (2.12)$$

where its frequency ω equals the plasma frequency ω_{p}. For $\omega_{\text{p}} > \omega$ ($n_e > n_c$) the refractive index becomes imaginary and the wave will be reflected. A plasma is therefore regarded underdense or overdense with respect to this critical density, allowing or inhibiting electromagnetic wave propagation, respectively. For a laser at wavelength $\lambda = 795$ nm the critical density amounts to $n_c = 1.7 \times 10^{21}$ cm^{-3}.

2.1.2. Nonlinear propagation, self-focusing and channel formation

The refractive index of a plasma as given in Eq. (2.11) only holds for the propagation of electromagnetic waves at low intensities. For laser pulses at relativistic intensities the electrons will oscillate with velocities close to the speed of light, as shown above. Therefore, the relativistic mass increase γm_e will modify the refractive index,

$$\eta(r) = \sqrt{1 - \omega_p^2(r)/\omega^2}, \qquad \omega_p(r) = \sqrt{\frac{n_e(r)e^2}{\gamma(r)m_e\varepsilon_0}} \, . \tag{2.13}$$

The relativistic modification of the refractive index is related to the laser intensity via $\gamma(r) \simeq \sqrt{1 + a^2(r)/2}$, where $a^2(r) \sim I(r)$. For a laser intensity peaked on axis $\partial I/\partial r \leq 0$, e. g. a gaussian profile, the refractive index is higher on axis and decreases with distance from the axis $\partial \eta/\partial r < 0$, which represents a positive lens. Owing to the radially modified index of refraction the laser will undergo relativistic self-focusing (RSF), if its power exceeds a critical power [44–46] $P > P_{\mathrm{RSF}}$ with

$$P_{\mathrm{RSF}} = 2 \left(\frac{m_e c^2}{e} \right) \left(\frac{4\pi\varepsilon_0 m_e c^3}{e} \right) \left(\frac{\omega}{\omega_{p,0}} \right)^2 \overset{(2.9)}{\simeq} 17.4 \ \mathrm{GW} \times \frac{n_c}{n_e}, \tag{2.14}$$

where $\omega_{p,0}$ is the unmodified plasma frequency.

An intuitive estimate for this power threshold can be obtained from geometrical treatment of diffraction and self-focusing for a Gaussian laser beam [35]. A laser beam with a Gaussian profile

$$a(r) = a_0 \exp\left[-\frac{r^2}{2\,w_0{}^2} \right] , \tag{2.15}$$

given in terms of the dimensionless amplitude $a(r)$, will be focused to a spot size w_0 and diffract with a divergence angle

$$\theta = \frac{dR}{dZ} = \frac{w_0}{Z_{\mathrm{R}}} = \frac{\lambda}{2\pi w_0} = \frac{c}{\omega\, w_0}, \tag{2.16}$$

where $Z_R = 2\pi\,w_0{}^2/\lambda$ is the Rayleigh length. Considering Eq. (2.13) the refractive

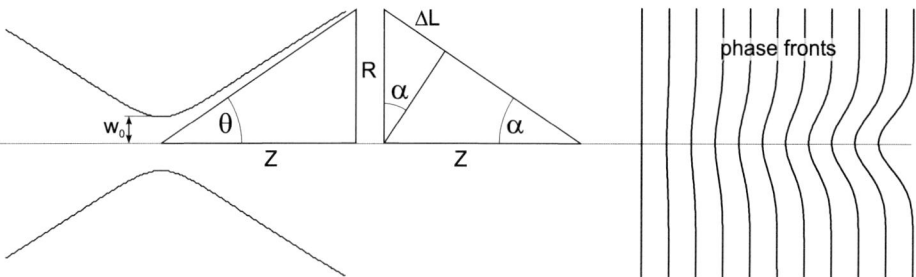

Figure 2.2: Relativistic self-focusing. The focusing angle from nonlinear interaction of the laser pulse with the plasma compensates for the gaussian beam diffraction, $\alpha \overset{!}{=} \theta$. Reproduced from [35].

index for this beam can be expressed as

$$\eta(r) = \sqrt{1 - \frac{\omega_{p,0}^2}{\omega^2 \left[1 + a^2(r)/2\right]^{1/2}}} \, , \tag{2.17}$$

The wave front of the beam propagating through this plasma will have a radially dependent phase velocity according to the refractive index $\eta(r)$. This phase velocity can be approximated by expanding both square roots:

$$\frac{v_{\text{ph}}(r)}{c} = \frac{1}{\eta} \approx 1 + \frac{\omega_{p,0}^2}{2\omega^2} \left[1 - \frac{a^2(r)}{4}\right] \, , \qquad \frac{1}{\sqrt{1\pm x}} = 1 \mp \frac{x}{2} + \dots \tag{2.18}$$

The phase front will travel off axis at the phase velocity in the ambient plasma at the edge, while their velocity will be decreased on axis according to the relativistic effect. The velocity difference will be

$$\frac{\Delta v_{\text{ph}}(r)}{c} = \frac{\omega_{p,0}^2}{8\omega^2} a_0^2 \exp\left[-\frac{r^2}{w_0^2}\right] \, . \tag{2.19}$$

The resulting curvature of the phase front will lead to focusing. The rays along the phase front will bend according to their path difference. The maximum path difference between the center and the edge is

$$\Delta L = [\Delta v_{\text{ph}}]_{\text{max}} \, t = \left[\frac{\Delta v_{\text{ph}}(r)}{c}\right]_{\text{max}} Z = \frac{\omega_{p,0}^2 a_0^2}{8\omega^2} Z = \alpha R \, . \tag{2.20}$$

9

The maximum angle resulting from this plasma focusing of the beam is given by

$$\alpha^2 = \omega_{p,0}{}^2 a_0{}^2 / 8\omega^2 , \qquad (2.21)$$

and therefore, depends on the intensity of the laser pulse. The divergence of the beam due to diffraction will be compensated by this self-focusing if $\alpha \geq \theta$ (Fig. 2.2). Comparison of Eq. (2.21) and Eq. (2.16) yields the power threshold for relativistic self-focusing as stated in Eq. (2.14)

$$a_0{}^2 w_0{}^2 \geq \frac{8c^2}{\omega_p{}^2} \quad \xrightarrow{P = I_0 A_0 \,\sim\, a_0{}^2 w_0{}^2} \quad P \geq \underbrace{\frac{8\pi\varepsilon_0 m_e{}^2 c^5}{e^2}}_{17.4\ \text{GW}} \frac{\omega^2}{\omega_{p,0}{}^2} . \qquad (2.22)$$

For our laser at $\lambda = 795$ nm propagating in plasma densities between $n_e \sim 10^{18} \ldots 10^{20}$ cm^{-3} the critical power for relativistic self-focusing to occur amounts to $P_{\text{RSF}} \sim 30 \ldots 0.3$ TW.

Self-focusing is just one example for nonlinear propagation of electromagnetic waves in a plasma and has been investigated theoretically by Esarey *et al.* [47] and experimentally by Fedosejevs *et al.* and Sarkisov *et al.* [48, 49]. For the interpretation of the experiments further nonlinearities arising from high intensity laser plasma interaction have to be discussed, such as wakefield generation (Sec. 2.2). But beforehand, the processes of channelling and cavitation have to be mentioned, which are closely related and intertwined with relativistic self-focusing discussed above.

A very interesting and important entity when dealing with the actions of real laser pulses on plasmas is the ponderomotive force [35, 43, 50]. For real laser pulses the electromagnetic waves are no longer considered to be an infinitely extended plane wave but rather exhibit a finite temporal and spatial envelope. The effect of this envelope becomes very apparent, if the equation of motion of an electron is averaged over the fast laser oscillations,

$$\vec{F}_{\text{pond}} = -\frac{e^2}{4\langle\gamma\rangle m\omega^2} \text{grad}\left[\vec{E}^2(\vec{r})\right] . \qquad (2.23)$$

Electrons that are subjected to a laser pulse feel this ponderomotive force, that is driving them away from high-intensity regions. Naturally, this force is most prominent in the focus of a gaussian laser pulse, where the highest intensities are reached.

The radially modified refractive index (2.13) also depends on the plasma den-

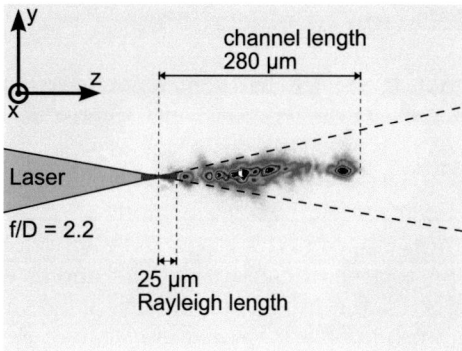

Figure 2.3: Self-focusing and channelling. A JETI laser pulse focused with a ratio of $f/D = 2.2$ will have a Rayleigh length of 25 μm in vacuum. Due to relativistic self-focusing and channelling in a plasma the beam diffraction may be compensated and the laser pulse is generating a relativistic channel guiding the pulse over more than ten times its Rayleigh length. The emission from nonlinear Thomson-scattering from the channel depends on laser intensity and electron density, which may differ along the channel due to self-focusing and cavitation.

sity profile $n_e(r)$, which, in turn, is modified by the ponderomotive expulsion of electrons from the beam axis. This effect enhances the self-focusing process and leads to the formation of a relativistic plasma channel. Including both relativistic self-focusing and ponderomotive self-channelling detailed studies [46] find that the power threshold for optical guiding in the plasma is then given by $P \geq 16.2$ GW $\times (n_c/n_e)$. In this self-channelling process the laser pulse expels electrons from the axis, leaving the ions behind. Analyzing the balance between the Coulomb force resulting from the charge separation and the ponderomotive force from the laser gives the cavitation condition [35]

$$\left(\frac{I \times (\lambda[\mu\text{m}])^2}{10^{19} \text{ W/cm}^2} \right) > \frac{1}{20} \left(\frac{n_e}{10^{19} \text{ cm}^{-3}} \right) \left(\frac{w_0}{\mu\text{m}} \right)^2 , \tag{2.24}$$

which denotes the intensity necessary to fully evacuate the region on axis from electrons. An experimental example for self-focusing, channeling and cavitation is given in Fig. 2.3.

2.2. Laser acceleration of electrons

In the preceding section it was discussed how nonlinear propagation of a laser pulse through a plasma affects the transverse characteristics of the pulse. For laser electron acceleration one has to take a closer look, now, at what the highly intense laser pulses are doing to the plasma. It has been calculated by Tajima and Dawson [51] that intense electromagnetic pulses are prone to create a wake of plasma oscillations and electrons trapped in this wake can be accelerated. A good overview of wakefield accelerators is given by Esarey *et al.* [52] and by Krushelnick *et al.* [53] who summarize the progress towards the production of monoenergetic beams. The electron energies up to 300 MeV [54] have been achieved with laser accelerators.

2.2.1. Laser wakefield acceleration

An intense laser pulse propagating in an underdense plasma pushes electrons back and forth by its ponderomotive force and leaves behind a plasma wave – also known as plasma wake [51]. Electron sheaths oscillate longitudinally with respect to the immobile ion background. The phase velocity of this plasma wake equals the group velocity of the driving laser pulse,

$$v_\mathrm{p} = \omega_\mathrm{p}/k_\mathrm{p} = v_\mathrm{L,g} = \eta\, c\,, \tag{2.25}$$

where η is the refractive index of the plasma from Eq. (2.11). Wakefield generation in a plasma works most efficiently if the laser pulse length matches half the plasma wavelength,

$$c\tau = \lambda_\mathrm{p}/2 = \pi c/\omega_\mathrm{p}\,. \tag{2.26}$$

For plasmas of density $n_\mathrm{e} \sim 10^{18}\ldots 10^{20}$ cm^{-3} this condition can be met by laser pulses with durations $\tau \sim 56\ldots 5.6$ fs, which is achievable by state-of-the-art laser systems. For such short pulses at moderate intensities ($a_0 \approx 1$) the wakefield will comprise several periods. In section 2.2.3 the case of higher intensities ($a_0 \gg 1$) will be considered, where the wake consists only of a single cavity.

Electrons that are injected into or trapped in such a laser wakefield can "surf" the wake to gain energy. The maximum accelerating electric field in a plasma may be approximated by [55]

$$E_\mathrm{p} = \frac{m_\mathrm{e} c\, \omega_\mathrm{p}}{e} \overset{(2.9)}{\simeq} \left(\frac{n_\mathrm{e}}{\mathrm{cm}^{-3}}\right)^{1/2} \mathrm{V/cm} \tag{2.27}$$

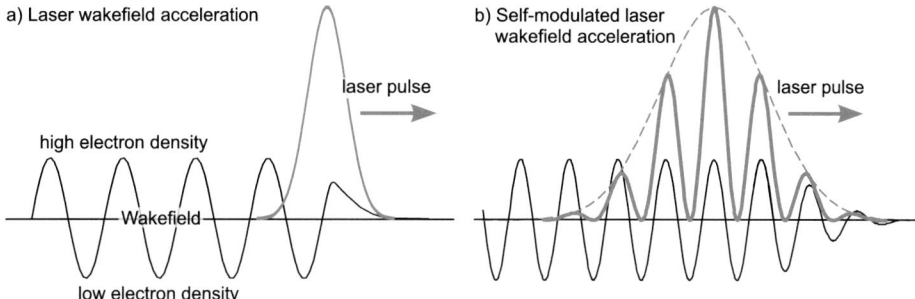

Figure 2.4: (a) Laser wakefield acceleration (LWFA). The laser pulse propagates through the plasma and pushes electrons back and forth by its ponderomotive force. This initiates an electron oscillation with respect to the immobile ion background that is driven by the space charge separation. Electrons injected in this wakefield may gain energy in the field between high and low electron density. (b) Self-modulated laser wakefield acceleration (SMLWFA). For laser pulses longer than the plasma wavelength the induced wakefield is modulating the pulse envelope, because for low density regions the group velocity will be higher than for high density regions. The laser pulse breaks up into a train of shorter pulses.

Thus, in a plasma of density $n_\mathrm{e} = 10^{20}$ cm^{-3} electric fields of the order of $E_\mathrm{p} \simeq$ 10 GV/cm are theoretically possible.

The fields in such a wake cannot grow infinitely. If the longitudinal oscillation of electrons becomes so large, that the fluid velocity exceeds the phase velocity of the wake, then wave-breaking occurs. Sheets of electrons in the laser wakefield cross each other, i. e. the wave is losing its coherence. The process of wave-breaking provides the electrons for acceleration. Those electrons, which cause the wave to break, "overshoot" their oscillation, such that they do not feel the restoring force of the space charge separation anymore, and may be injected into the next wave bucket, where they will be accelerated.

If the laser pulses are longer than indicated by Eq. (2.26) the envelope of the laser pulse comprises several plasma periods. The density modulations lead to a variable refractive index and along the pulse. At regions where the electron density is low the group velocity will be higher than in regions with high electron density. Therefore, the pulse envelope will modulate which in turn enhances the wakefield generation, because the bunching of the envelope leads to an increased ponderomotive force (Fig. 2.4). This resonant process of mutual enhancement of plasma density modulation and envelope bunching will lead to a laser pulse consisting of several shorter pulses approaching the condition Eq. (2.26).

2.2.2. Direct laser acceleration

In relativistic channels in high plasma densities another electron acceleration process may occur. The channel produced by relativistic and ponderomotive channelling of the laser pulse generates a transverse electric field due to space charge separation between the pondermotively expelled electrons and the ions remaining in the channel. Therefore, the electrons oscillate in the combination of this quasi-static electric field and their self-generated magnetic field at the betatron frequency $\omega_\beta = \omega_p/\sqrt{4\gamma}$. In the case of resonance between the the betatron oscillation and the relativistic motion of the electron, it may obtain energy directly from the laser field. This regime, named direct laser acceleration (DLA), was identified experimentally and theoretically by Gahn *et al.* [56, 57] and Pukhov *et al.* [58].

2.2.3. Bubble acceleration

For high-intensity $(a_0 > 1)$ few-cycle laser pulses shorter than the plasma wavelength a new regime of laser wakefield acceleration – highly nonlinear broken wave regime – has been identified by Pukhov and Meyer-ter-Vehn [24]. In this regime, linear plasma theory is no longer sufficient to describe LWFA. Three-dimensional particle-in-cell (3D-PIC) simulations are the most widely used tool to investigate this highly non-linear evolution theoretically.

In simulations for ultra-relativistically intense laser pulses shorter than λ_p, it can be observed that the ponderomotive force of the laser creates a bubble-like electron void. Hence, this acceleration mechanism was dubbed *bubble acceleration*. The bubble comprises a fully broken plasma wave, supporting a strong longitudinal electric field. The laser pulse travels at the front of this structure and expels electrons which stream around the arising cavity and can be trapped inside at a fixed phase of the longitudinal field, leading to the acceleration of electron bunches with a narrow energy spread.

This phenomenon has been thoroughly explored theoretically [59–61]. Owing to the large amount of numerical simulations and a similarity theory, laser and plasma parameters required for bubble acceleration to occur have been identified. For electron bunches generated within this parameter space, electron energy and numbers can be calculated by scaling laws. Optimum conditions for bubble acceleration in a plasma with wave number $k_p = \omega_p/c$ are expected for laser pulses with

$$k_{\mathrm{p}}R \approx \sqrt{a_0} \,, \qquad \tau \leq \frac{R}{c}, \tag{2.28}$$

for focal spot radius R and pulse duration τ, respectively. From those conditions a threshold power to reach the bubble regime may be derived:

$$P > P_{\mathrm{crit}} = P_{\mathrm{rel}}\,(\omega_0\tau)^2 \approx \left(\frac{\tau\,[\mathrm{fs}]}{\lambda\,[\mu\mathrm{m}]}\right)^2 \times 30\,\mathrm{GW}\,, \tag{2.29}$$

where $P_{\mathrm{rel}} = m_e^2 c^5/e^2 \approx 8.5\,\mathrm{GW}$ is the natural relativistic power unit and τ and λ are the laser pulse duration and wavelength, respectively. Laser pulse power and duration determine the energy of the monoenergetic peak in the electron spectrum. This scaling is given by

$$E_{\mathrm{mono}} \approx 0.65\,m_e c^2 \sqrt{\frac{P}{P_{\mathrm{rel}}}}\frac{c\tau}{\lambda} = 0.1\,\mathrm{MeV} \times \sqrt{\frac{P}{P_{\mathrm{rel}}}} \times \frac{\tau\,[\mathrm{fs}]}{\lambda\,[\mu\mathrm{m}]}\,. \tag{2.30}$$

The number of electrons in such an electron bunch also depends on the square-root of the laser pulse power,

$$N_{\mathrm{mono}} \approx \frac{1.8}{k_0 r_e}\sqrt{\frac{P}{P_{\mathrm{rel}}}} = 10^8 \times \sqrt{\frac{P}{P_{\mathrm{rel}}}} \times \lambda\,[\mu\mathrm{m}]\,, \tag{2.31}$$

where k_0 is the laser wave number and $r_e = e^2/4\pi\varepsilon_0 m_e c^2$ is the classical electron radius. The conversion efficiency of laser energy into electrons in the monoenergetic bunch is constant: $(N_{\mathrm{mono}}E_{\mathrm{mono}})/(P\tau) \approx 20\,\%$, suggesting the bubble regime as a promising laser plasma acceleration scenario.

2.3. Laser acceleration of protons and ions

Proton or ions cannot be accelerated by the laser directly, at least not at intensities currently available. Therefore, the proton acceleration always relies on an electron acceleration by the laser beforehand. The most efficient scenario for proton acceleration is the irradiation of thin metal foils by an high intensity laser pulse [3, 4, 6]. The world record for laser accelerated protons lies at an energy of 58 MeV, produced already in the first experiments using the large-scale petawatt laser at Lawrence Livermore National Laboratory, USA [6, 62]. Further investigation of ion acceleration indicated that laser produced ion beams promise ultra-low emittance both transversely as well as longitudinally [28].

2.3.1. Laser plasma interaction with solid targets

Due to the index of refraction of a plasma Eq. (2.11), which depends on the plasma electron density, laser pulses cannot propagate at plasma densities above the critical density. Therefore, the electron acceleration and plasma heating have to occur before and directly at the plasma boundary given by the critical density.

Resonance absorption

Highly amplified laser pulses exhibit a pedestal of amplified spontaneous emission (ASE) preceding the main laser pulse by several nanoseconds. The intensity of the ASE is several orders of magnitude lower than the one of the laser pulse, but nevertheless sufficient to ionize material before the actual high-intensity laser pulse is arriving. A pre-plasma will be generated that is expanding in advance of the main laser pulse. An obliquely incident laser pulse exhibits a target normal component of its electric field. This field component couples to the collective plasma motion. As the laser pulse is propagating towards the increasing density this coupling becomes more efficient, and near the critical density, where the plasma frequency equals the laser frequency, the case of resonance absorption is achieved [35, 43, 63]. Resonance absorption is effective for intensities below the relativistic limit and yields "hot" electron distribution with temperatures of several $10\,\text{keV}$. For relativistic intensities the $\vec{v} \times \vec{B}$ heating becomes dominant [63].

Ponderomotive acceleration and $\vec{v} \times \vec{B}$ heating

The ponderomotive force of a laser pulse Eq. (2.23) is pushing the electrons away from intensity gradients. The electrons in the plasma in front of the target will be accelerated with a nearly thermal distribution, which can be associated with a temperature that may be obtained from the ponderomotive potential [63, 64],

$$
\begin{aligned}
\Phi_{\text{pond}} &= m_e c^2 (\gamma - 1) \\
&= m_e c^2 \left(\sqrt{1 + a_0{}^2/2} - 1 \right) \quad\quad\quad (2.32) \\
T_{\text{e,pond}} &\simeq 0.511\,\text{MeV} \times \left(\sqrt{1 + a_0{}^2/2} - 1 \right), \quad\quad (2.33)
\end{aligned}
$$

because in the case of linearly polarized laser light the relativistic factor is given by $\gamma = \sqrt{1 + a_0{}^2/2}$. For intensities $I_\text{L} = 10^{18} \dots 10^{20}\,\text{W/cm}^2$ temperatures in the range of $T_{e,pond} \simeq 0.1 \ldots 2.0\,\mathrm{MeV}$ result from this scaling. In reality the absorption of laser light and the transfer of energy to hot electrons will be a combination of resonance absorption, vacuum heating and $\vec{v} \times \vec{B}$ heating.

In the heated blow-off plasma at the front surface the electrons will be expelled, whereas the protons will stay behind due to their higher inertia. At a later stage a combination of this space charge separation, and the coulomb explosion of the ion distribution, will provide for the proton and ion acceleration from the front of the target [65]. However, the laser will drive the electrons predominantly into the target. If the target is thin enough, i. e. if it is a foil, the electrons will penetrate the foil and leave it behind with a positive charge. Between this space charge separation at the target back side an electrostatic field builds up that can be sustained long enough to accelerate protons in the forward direction. Because the field at the target back side by an electron sheath that has been driven through the target and will act perpendicularly to the target surface this mechanism for proton acceleration was named Target Normal Sheath Acceleration (TNSA) [66]. The protons have been identified to originate from a contamination layer of water vapor and carbohydrates on the metal surface [67]. The generation of proton beams from the target back side has been observed in numerous experiments and was found to surpass the front side acceleration with respect to efficiency, energy and collimation of the produced particle beams [68–70].

2.3.2. Target Normal Sheath Acceleration

A one-dimensional fluid model of TNSA

The acceleration process on the back side of a foil can be described as a fast rarefaction of an electron and ion distribution due to expansion of a plasma into vacuum. This model was proposed by Mora [71] and has been very successful in the theoretical description of proton acceleration experiments [72–74]. For $t = 0$ the ion distribution exhibits a sharp density gradient, described by a step function from $n_i(z \leq 0, t = 0) = Z n_{e,0}$ to $n_i(z \geq 0, t = 0) = 0$. The hot electrons are assumed to be Boltzmann distributed in equilibrium to the electrostatic potential between electron and ion distribution:

$$n_e(z, t) = n_{e,0} \exp\left[\frac{e\,\phi(z, t)}{T_{e,h}}\right] \tag{2.34}$$

The electrostatic potential obeys Poisson's equation for the given charge distribution,

$$\varepsilon_0 \frac{\partial^2 \phi(z,t)}{\partial z^2} = e\left[n_e(z,t) - n_i(z,t)\right] . \tag{2.35}$$

Solving Eq. (2.35) the initial electric field can be calculated

$$\mathcal{E}_0 = -\left.\frac{\partial \phi(z,t=0)}{\partial z}\right|_{x=0} \approx \sqrt{\frac{T_{e,h} n_{e,0}}{\varepsilon_0}} . \tag{2.36}$$

For a hot electron temperature $T_{e,h} = 1\,\mathrm{MeV}$ and an electron density of $n_{e,0} = 10^{20}\,\mathrm{cm}^{-3}$ the initial field is about $\mathcal{E}_0 \approx 10^{10}\,\mathrm{V/cm}$. The expansion into vacuum for $t > 0$ is then described via the continuity equation and the momentum equation,

$$\left[\frac{\partial}{\partial t} + v_i(z,t)\frac{\partial}{\partial z}\right] n_i(z,t) = 0$$

$$\left[\frac{\partial}{\partial t} + v_i(z,t)\frac{\partial}{\partial z}\right] v_i(z,t) = -\frac{e}{m_i}\frac{\partial \phi(z,t)}{\partial z} , \tag{2.37}$$

where $v_i(z,t)$ is the ion velocity.

From this model one can obtain two formulas describing the energy spectrum of the highest energy protons:

$$\frac{dN}{dE} = \frac{n_{e,h} A_{source}\, c_s t}{Z\sqrt{2ZT_{e,h}E}} \exp\left[-\sqrt{\frac{2E}{ZT_{e,h}}}\right] , \tag{2.38}$$

and the maximum proton energy:

$$E_{max} = 2ZT_{e,h}\left[\ln\left(\tau + \sqrt{\tau^2 + 1}\right)\right]^2 . \tag{2.39}$$

In both formulas, Z is the ion charge, E is the proton energy, $c_s = \sqrt{ZT/m_i}$ is the ion sound speed. Furthermore, the acceleration time is normalized to $\tau = \omega_{p,i} t/\sqrt{2e}$, with $e = 2.718\ldots$, and $\omega_{p,i} = \sqrt{Zn_{e,h}e^2/m_i\varepsilon_0}$ the ion plasma frequency known from Eq. (2.9). In order to get the proton number also the area A_{source} of the rear-side electron spot has to be taken into account.

With a set of values for our experimental conditions one can give an estimate for the proton spectra and cutoff energies to be expected. The hot electron temperature may be taken from the ponderomotive scaling from Eq. (2.32),

$T_{e,h}(a_0 = 4) \simeq 1.0\,\mathrm{MeV}$. Following the considerations of Kaluza *et al.* [73], the laser energy E_L transferred to hot electrons $T_{e,h}$ with a conversion efficiency η_e for a known focus size w_0 gives the electron density $n_{e,h}$ at the target back side,

$$n_{e,h} = \frac{\eta_e E_L / T_{e,h}}{c\,\tau_L\,\pi(w_0 + d'\tan\vartheta_{in})}\,, \qquad (2.40)$$

where $d' = d/\cos 45°$ is the effective target thickness and ϑ_{in} is half the opening angle of the electron transport through the target. Therefore, all numbers required to evaluate equations (2.38) and (2.39) can be estimated from the laser and target conditions. The proton spectrum and cutoff energy depend on the acceleration time, which should be similar to the duration of the laser pulse, which is driving the hot electrons. Fuchs *et al.* [72] found that using this model, experimental results could be reconstructed best, if the acceleration time is assumed to be $t_{acc} \simeq 1.3\,\tau_L$. The spectral evolution for our laser parameters is shown in Fig. 2.5. At an acceleration time of 104 fs (1.3×80 fs) this results in a quasi exponential spectrum with a cutoff energy around $4\,\mathrm{MeV}$.

The estimation for proton acceleration as described above is legitimate, since results of many experiments may be predicted or analyzed accurately based on this model. But the numbers expected from this model have to be used advisedly, because, first of all, the expansion was described as an isothermal process, which is clearly a simplifying assumption for a fast expanding plasma, and furthermore, many of the parameters involved, such as the conversion efficiency of laser energy into electrons (30 % [75], 40 . . . 50 % [62]), the electron density at the back side, and the acceleration time, are rather vague, because they cannot be measured directly.

Furthermore, the fluid model was restricted to one dimension for the sake of simplicity. The real situation has to be addressed under consideration of three dimensions, and is very sensitive to the extension and the shape of the accelerating field at the target backside. Not only the energetic electron distribution behind the target determines the characteristics of the field, but also the spatial electron distribution plays a major roll in the generation of proton beams from TNSA.

Predictions from numerical simulations

In order to get a more reliable and quantitative description of the acceleration one has to resort to particle-in-cell or tree code simulations [66, 76–82]. Such simulations give a very accurate account of the mechanism including more parameters and dimensions as the previously mentioned fluid model. Therefore, experimental

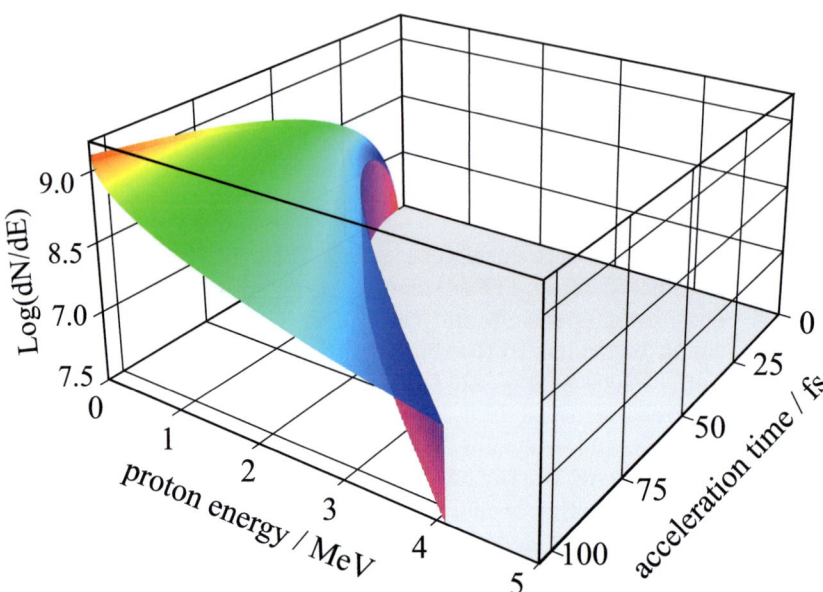

Figure 2.5: Estimation of a proton spectrum for TNSA for laser irradiation $a_0 = 4$, based on Eq. (2.38) for acceleration times $0\ldots104$ fs. The cutoff energy for the spectra is calculated from Eq. (2.39).

results may be interpreted and verified by numerical models. In section 4.4 this interplay between experiment and simulation will be demonstrated for the case of our results on proton acceleration from microstructured targets, which were corroborated by simulations performed by Timur Esirkepov from JAERI in Japan. The work of Esirkepov *et al.* [78] also presents the foundation and stimulus for the experiments presented in section 4.3. They simulated the TNSA field in three dimensions, and found that the TNSA field has cylindrical symmetry and can be assumed nearly homogeneous in the range defined by the transverse extension of the electron distribution. The field extends transversely over a few hundred µm and decreases rapidly in transverse direction outside the range defined by the laser focus, giving rise to the broad quasi exponential proton/ion spectra with 100 % energy spread. This statement is supported experimentally by measurements of the source size of the proton/ion emission [83–85].

Therefore, it was proposed to exploit the nearly homogeneous region of the

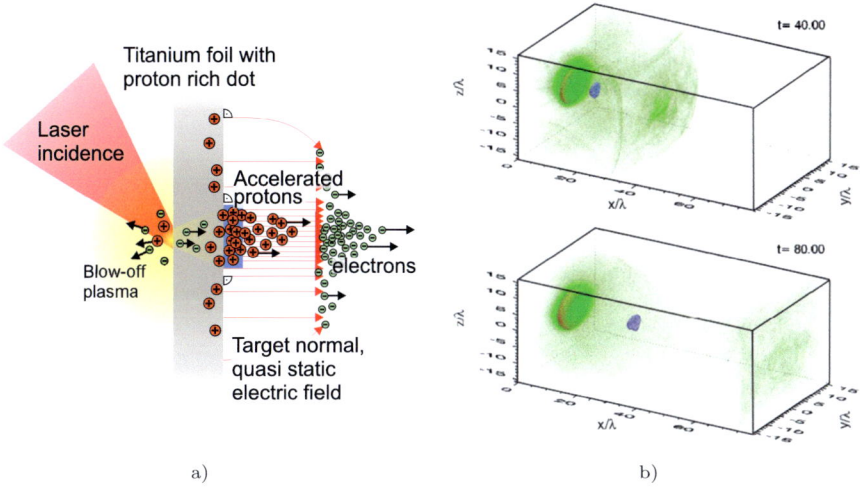

Figure 2.6: Target Normal Sheath Acceleration (TNSA) of protons from the back side of a micro structured target. (a) A TW-laser pulse is focused onto the front side of the target foil, where it generates a plasma and subsequently accelerates electrons. The electrons penetrate the foil, ionize hydrogen and other atoms at the back surface and set up a Debye sheath. The inhomogeneous electron distribution in the sheath causes a transversely inhomogeneous accelerating field. Applying a small hydrogen rich dot on the back surface enhances the proton yield in the central part of the accelerating field, where it is nearly homogenous. These protons constitute the quasi mono-energetic bunch. (b) Figure taken from [78]. Two time steps for the simulation of the acceleration of a micro-dot from the target back side are displayed in a density plot. Green is the electron distribution, red are the heavy ions from the target foil, whereas the protons from the dot are represented by purple.

field in the range determined by the laser focus by applying a second layer of a hydrogen rich material on the target back side, which matches the dimensions of the homogeneous region. Such a *dot* should lead to an increased proton yield from that area. In their simulation a uniform acceleration leading to monoenergetic protons was observed consequently. An impression of this *confined* TNSA may be received from Fig. 2.6. The left sketch shows the combination of TNSA and the described modified target in an intuitive way. The laser pulse is impinging on the foil and accelerating electrons, which then will lead to the accelerating Debye sheath at the target back side. By modifying the target in such a way that the suitable part of the field is used only, the bandwidth of the proton energy spectra may be reduced. Fig. 2.6b shows this very situation observed in the simulations published in [78]. The electrons (green) are expelled from the target back side, whereas the heavy ions (red) and protons (purple) will follow the accelerating field later. The proton dot detaches as a whole from the heavy ion layer and is accelerated in a bunched manner.

§ 3 Electron Acceleration

In this chapter the laser plasma accelerator for electrons will be presented. Laser induced electron acceleration relies on the interaction of a laser pulse with a plasma in order to generate plasma waves which support high gradients in longitudinal direction that allow for a high energy gain of electrons trapped inside those plasma waves. Electrons with energies of more than $300\,\mathrm{MeV}$ have been observed from laser plasma accelerators [54]. Also a new regime of laser electron acceleration – bubble acceleration – has been identified [24] and lead to the generation of quasi-monoenergetic electron bunches [21–23, 53, 86–88].

A laser plasma accelerator operating with longer laser pulses and higher plasma densities was demonstrated, which is as well capable of producing electron bunches with greatly reduced energy spread and a high level of collimation (Sec. 3.5) [26]. Furthermore, the interaction leading to the relativistic channels being the source of the high energy electrons was investigated, and the first optical observation of the electron bunch inside such an accelerating bubble will be described (Sec. 3.3 and 3.6) [27].

But before the experiment itself will be discussed, the laser system, which is delivering the ultra-short high-intensity pulses for our experiments, will be introduced.

3.1. JETI – The Jena Titanium:Sapphire Laser

All of the experiments presented here were performed at the multi-TW JETI laser system [89, 90]. This laser applies the principle of chirped pulse amplification (CPA) [91–93] to generate pulses of ultrahigh intensities. In Fig. 3.1 a schematic overview of the JETI is given. An oscillator, pumped by a frequency doubled Neodymium:YVO$_4$ laser generates laser pulses of 45 fs duration at a center wavelength of $\lambda_0 = 795$ nm with a repetition rate of $80\,\mathrm{MHz}$. Each pulse contains an energy of 10 nJ. It is technologically not possible to amplify all pulses at such a high repetition rate. Therefore, pulses at a rate of 10 Hz are picked from the pulse train by a pockels-cell–polarizer combination. A grating stretcher introduces a positive chirp ($d\lambda/dt < 0$) to the pulse which is therefore stretched to a pulse duration of 150 ps.

The amplification of the stretched pulses comprises three stages which are all pumped by frequency-doubled Neodym:YAG lasers. The lion's share of amplification is delivered by approximately 20 roundtrips in a regenerative amplifier yielding pulses at about 2 mJ. Consecutive amplification in a 4-pass and a 3-pass amplifier result in pulse energies in the range of $1 \ldots 1.5$ J, which represents an overall amplification factor of 10^8.

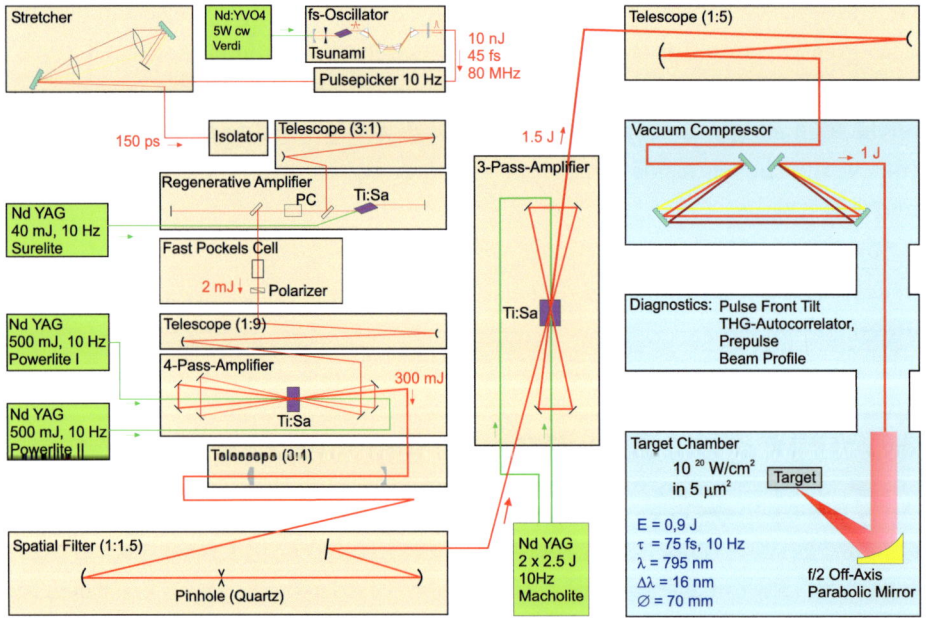

Figure 3.1: JETI: Schematic of the Jena Titanium:Sapphire laser system. Chirped pulse amplification increases the laser energy within three amplification stages. Pulse compression, diagnostics and experiments take place in vacuum.

The fully amplified pulses are recompressed to about 80 fs in a 4-grating compressor. The initial pulse duration can not be achieved anymore due to gain narrowing in the amplification process and the introduction of higher order chirp in the system. The energy transmission through the compressor amounts to about 60 %, resulting in pulses with an energy between $0.6 \ldots 0.9$ J available for experiments at a power of > 10 TW.

This stretching, amplification and consecutive compression is the essential concept of CPA. Intensities are kept below thresholds for damage of optical components as well as occurrence of non-linear interaction during the propagation in air, because the pulses are stretched and the beam diameter is enlarged several times by telescopes in the laser system. The final beam diameter is approximately 50 mm. The grating compressor and the consecutive mirrors for guiding the high power pulses to the interaction chamber are all operated in vacuum.

For the proton acceleration experiments presented in Sec. 4 the laser system was equipped with an additional fast pockels cell in between the regenerative amplifier and the 4-pass [94]. The pockels cell and its high voltage driver have rise times of 250 ps and 150 ps, respectively. With variable delay between the pockels cell driver and the main laser pulse the ASE level preceding the laser pulse may be suppressed by a factor of 10^{-3} in a range of $0.5 \ldots 5$ ns before the main pulse. The temporal structure of the laser pulse is displayed in the Fig. 3.2. With this additional pockels cell the ASE level may be reduced to 10^{-9} compared to the main pulse. In the nanosecond range before the pulse the temporal structure was determined using a fast photo diode, whereas the picosecond vicinity of the pulse was measured with a background-free THG autocorrelator.

3.2. Experimental setup

The setup for the experiments on electron acceleration is situated in a vacuum chamber and a schematic is displayed in Fig. 3.3. The laser pulses provided by JETI, as described in the preceding section, were focused by a gold coated off-

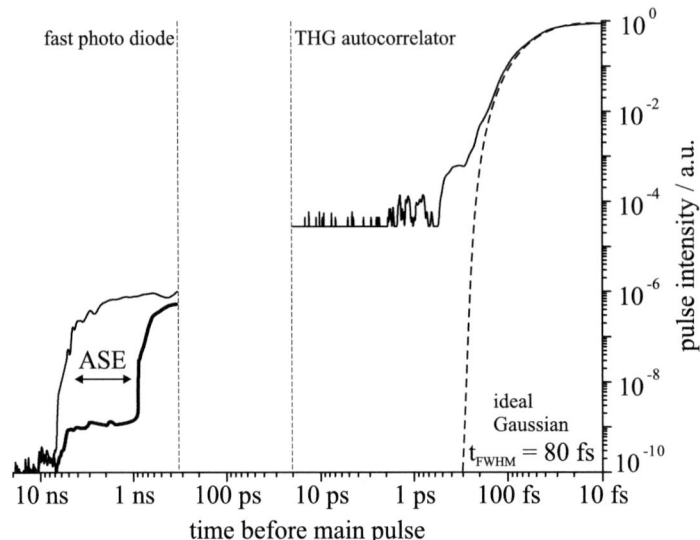

Figure 3.2: Temporal profile of a JETI pulse. The temporal profile up to 15 ps before the main pulse was determined by a background-free THG autocorrelator. The ASE level was measured for times greater than 0.5 ns using a fast photodiode. With the additional pockels cell the ASE level could be reduced by three orders of magnitude in the range of 0.5 ... 5 ns before the main pulse.

axis parabolic mirror with a focusing ratio of $f/D = 2.2$, before they reach the target for interaction. This concentrates the terawatt pulses to an area of about $5 \ldots 8 \ \mu m^2$ yielding intensities of $2 \times 10^{19} \ldots 5 \times 10^{19}$ W/cm^2. The intensities for the experiment are measured indirectly by measuring pulse energy, pulse duration and the focal spot area. Consisting of several measurements the intensity may be given with an accuracy of 30%. A very detailed description about the measurement of intensities is given in [89]. The axis of propagation of the main laser pulse (red pulse in Fig. 3.3) is defined to be along the z-axis. The laser polarization is along the x-axis.

The main laser pulse was focused onto a pulsed He gas jet, which delivered the target material for plasma generation, self-focusing and self-channelling, and consequently for electron acceleration as described in the theory sections Sec. 2.1.2

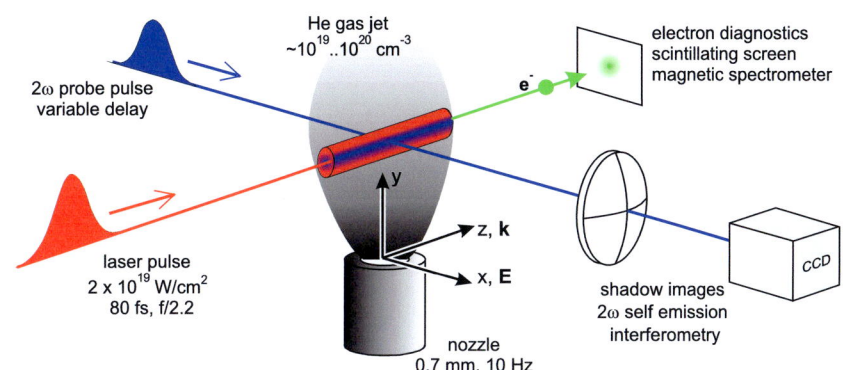

Figure 3.3: Experimental setup: The main pulse is focused to an intensity of a few 10^{19} W/cm^2 into a pulsed He-gas jet, where a relativistic channel due to relativistic self-focusing is forming. By splitting and frequency doubling part of the main pulse a probe pulse is generated. This probe pulse is used for imaging and observing the laser- plasma interaction on a time scale determined by the probe pulse duration of ≈ 100 fs. The electrons that are accelerated in the relativistic channel are characterized by magnetic spectroscopy and nuclear reactions.

and Sec. 2.2. Helium was chosen because it is most suitable for self-focusing. It will be fully ionized by the rising edge of the laser pulse and ionization defocusing can be avoided [48, 49]. The gas flow from the nozzle was characterized, which is presented in the following section 3.2. The region above the nozzle, where the laser plasma interaction happened, was imaged onto a CCD camera placed in x-direction perpendicular to the main laser pulse. The imaging yielded many interesting observations both time-integrated and time-resolved, as will be explored in sections 3.3 and 3.6, respectively. For time-resolved measurements it is necessary to reach a time regime well below the one given by the exposure time of the camera. This was achieved by employing an ultra-short probe pulse as a backlight for the plasma. Section 3.5 addresses the characterization of the electron pulses that are accelerated from the laser plasma and their immediate application to induce nuclear reactions is explained later in section 5.2.

Gas targets for relativistic plasma physics

For performing experiments at a repetition rate of 10 Hz under constant conditions new target material has to be provided for each new laser shot. This was achieved by using a high speed solenoid valve (Parker Inc.) producing a pulsed He gas jet as a target. Because the relativistic interaction takes place over distances of a few hundred micrometers it is essential to know the gas density provided by the valve at the time of interaction. The produced gas density profile depends on the shape of the nozzle that is attached to the valve. The gas density was determined by measuring the refractive index of the gas by means of interferometry using a modified Nomarski-interferometer [95]. In the interferograms the integral phase shift $\Delta\Phi(z)$ was measured, where z is the coordinate in the interferogram. This phase shift results from accumulating the local phase shift $\Delta\phi(\rho) = k[\eta(\rho) - 1]$ by the interferometer beam with wavenumber k as it propagates through the gas with the distribution function of the refractive index $[\eta(\rho) - 1]$, with $\rho = \sqrt{x^2 + z^2}$ being the coordinate in the gas jet which is assumed to be cylindrical symmetric about the y-axis. For cylindrical symmetry $[\eta(\rho) - 1]$, may be inferred from the integral phase shift by Abel-inversion [96]. The gas pressure is then calculated from

$$\frac{p(z)}{1\,\text{bar}} = \frac{[\eta(\rho) - 1]}{[\eta - 1]_{1\,\text{bar}}},$$

by relating the local index of refraction to the refractive index of the gas at $p_{\text{atm}} = 1\,\text{bar}$. For a two dimensional interferogram this procedure is repeated for all coordinates y to give the two-dimensional pressure distribution (Fig. 3.4).

Because of the very small refractive index for helium $[\eta - 1]_{1\,\text{bar}} = 3.5 \times 10^{-5}$ and the small volume of the pulsed jet, nitrogen ($[\eta - 1]_{1\,\text{bar}} = 3.0 \times 10^{-4}$) or argon ($[\eta - 1]_{1\,\text{bar}} = 2.8 \times 10^{-4}$) was used instead to impose a measurable phase shift onto the interference beam. Argon was used to double check the results obtained with nitrogen, in order to assure that the density above the nozzle does not depend on the species of gas – wether it is mono- or diatomic.

A cylindrical capillary nozzle with an inner diameter of $D_{\text{i}} = 0.7$ mm has been

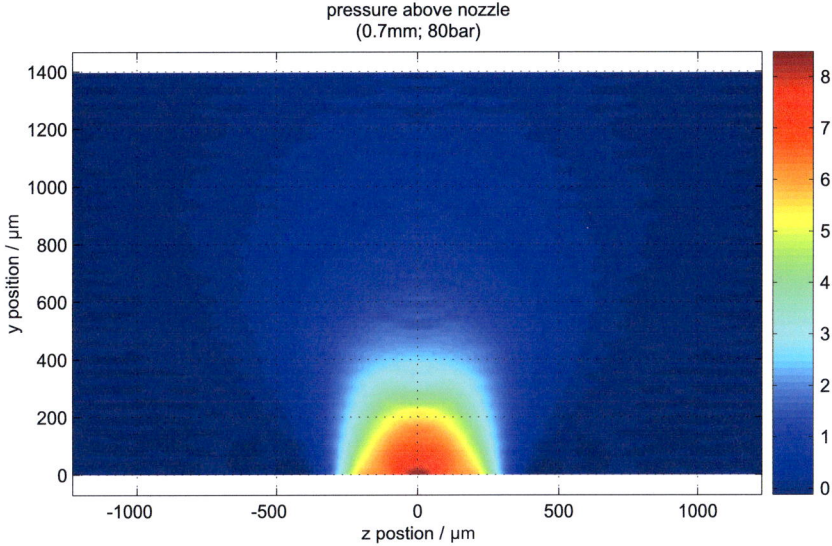

Figure 3.4: Pressure p above the nozzle. The colorbar to the right indicates the color associated to the pressures from 0 to 8 bar. Convert pressure to density using $n/\mathrm{cm}^{-3} = 2.4 \times 10^{19}\, p/\mathrm{bar}$. The point $(y = 0\,\mu\mathrm{m}, z = 0\,\mu\mathrm{m})$ marks the center of the orifice. There the maximum pressure of $\approx 8\,\mathrm{bar}$ is reached. The density shows a Gaussian profile laterally (z-direction) and decreases exponentially on axis.

used, which gives a gas pressure distribution as shown in Figure 3.4. The two-dimensional pressure distribution above the nozzle is represented by a color-code. At the center of the orifice the maximum pressure of $p_{\mathrm{max}} \approx 8\,\mathrm{bar}$ is reached. This valve was operated with backing pressure in the range of $p_{\mathrm{b}} = 10 \ldots 80\,\mathrm{bar}$. In this range the maximum pressure above the nozzle is linearly proportional to the backing pressure that is applied to the valve and depends on the opening time of the nozzle. From there the pressure decreases exponentially in y direction with increasing distance from the nozzle. The lateral profile has a Gaussian shape. Thus, the pressure above the nozzle can be approximated by the product of an exponential and a Gaussian function with cylindrical symmetry

$$p\left(y,\rho\right) = p_{\max}\, e^{-y/\mu}\, e^{-\rho^2/w^2(y)} \ . \tag{3.1}$$

These are the characteristics of a pressure profile for a subsonic flow from a nozzle. A pulsed laser was used to get time-resolved information about the gas pressure above the nozzle as the valve was switched by a fast high voltage pulse provided by the driver for the valve. The length of the switching pulse determined the opening time of the nozzle. The time-resolved measurements for different opening times show a linear increase in pressure till the respective opening time was reached and slower decrease afterwards attributed to the further expansion of the gas. The main laser pulse was consequently synchronized to the driving pulse of the nozzle, and fired at the moment of highest pressure.

The two-dimensional pressure distribution can be used as a mapping for the plasma density at the measured positions for the relativistic channels in the experiments (see Fig. 3.5). The pressure may be converted to gas density by

$$n_{\text{gas}} = \frac{n_{\text{e,He}}}{2} = 2.4 \times 10^{19}\,\text{cm}^{-3}\,\frac{p}{1\,\text{bar}} \ . \tag{3.2}$$

Since the laser intensity is high enough to fully ionize the helium, the plasma density n_{e} will be twice the gas density n_{gas}. For high intensities the leading edge of the laser pulse will ionize the gas and the main part of the pulse will interact with a plasma.

3.3. Channel observation

The measurement of the gas/plasma density in conjunction with relativistic channels observed via nonlinear Thomson scattering yields the density parameters necessary for optimum channel formation and electron acceleration in the case of the presented experiments.

The interaction region of laser and gas above the nozzle was imaged onto a CCD camera (Fig. 3.3). The observation direction was along the x-axis, perpendicular to the propagation direction of the main laser pulse. A lens $(f/D = 4.9)$ was placed at different distances to the plasma to change the magnification. The spatial resolution of the imaging amounts to $5 \ldots 10$ µm. The main pulse will undergo relativistic self-focusing in the plasma, generating a relativistic channel, as described in chapter 2.1.2. In this channel the electrons experience the large electromagnetic fields of the laser pulse at enhanced intensity. The laser pulse propagates in z-direction and is polarized along the x-axis and electron motion will result in the "Figure-8"-motion within the x-z-plane. The self-emission of the plasma in x-direction is spectrally close to the second harmonic of the laser, and is detected by the imaging lens and the camera. Since also higher harmonics result from nonlinear Thomson-scattering and stray light of the fundamental inside the chamber is present, a band pass filter at $\lambda = (400 \pm 5)$ nm is placed in front of the camera in order to observe light around the second harmonic only. Therefore, the 2ω-light is used to monitor the interaction and the relativistic channel is defined as the structure, which constitutes the source of this second harmonic emission, being a prominent feature of the relativistic nature of nonlinear Thomson scattering (Ch. 2). It was verified that the polarization direction of the second harmonic was in z-direction which is expected for second harmonic light generated by nonlinear Thomson scattering.

The positions and lengths of the channels with respect to nozzle were measured in the images and can be related to the respective plasma density (Equations (3.1), (3.2)) obtained from measurements presented above. Fig. 3.5 shows the lateral plasma density profile at the height above the nozzle where the laser focus is situated. The plasma density exhibits a gaussian profile and reaches its maximum at $z = 0$ above the center of the nozzle. The position of the relativistic channel is indicated by the inset, which shows a typical image of the Thomson light from the relativistic channel. Best results for channel length and electron acceleration were obtained, when the laser vacuum focus and the steepest gradient of the density coincided. Then, the channels extended to about $10 \ldots 12$ times the laser's

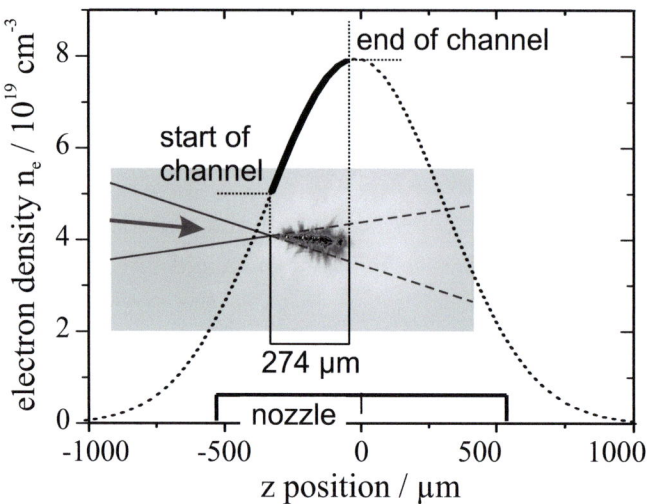

Figure 3.5: The lateral plasma density profile above the nozzle has a Gaussian shape (dotted line). The density along the channel and its position is indicated by the bold line. The inset shows the corresponding observed channel of the same spatial region. The extension of the emission indicates a channel length of 274 µm, about twelve times the Rayleigh length of the laser. The relativistic channel starts near the steepest density gradient and ends close to the maximum density (grad $n_e = 0$).

original Rayleigh length and strongly collimated electron jets from the channel were observed (Fig. 3.8).

At a plasma density of about $n_e \sim 5 \times 10^{19}$ cm^{-3}, where the relativistic channels start, the required power for relativistic self-focusing can be calculated by Eq. (2.14) to $P_{RSF} \sim 0.6$ TW. The laser power used in the experiments exceeds this requirement by a factor of ten. Also the cavitation condition Eq. (2.24) is fulfilled more than five times. Therefore, the emission intensity is not constant over the channel (cf. profile in Fig. 2.1) and decreases where most electrons are expelled from high laser intensity regions [97].

The channelling in the plasma was also observed by means of interferometry. The refractive index of a plasma is related to its electron density as given by Eq. (2.11). The phase shift imposed on a probe beam traversing a plasma may

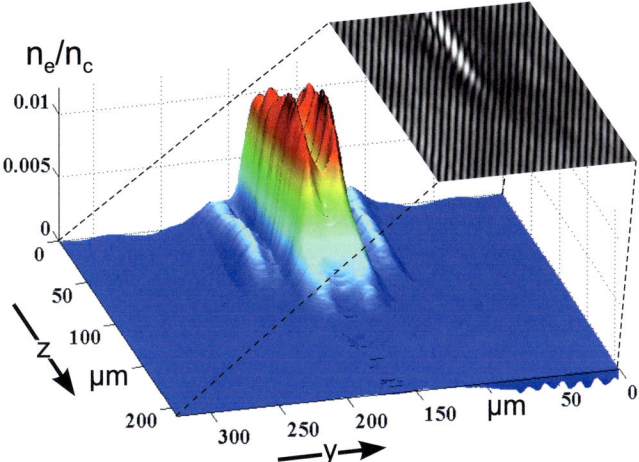

Figure 3.6: Plasma density at the start of the channel determined by interferometry. The electron density in the walls of the channel rose up to values of $n_{e,\text{wall}} = 6 \times 10^{19}$ cm^{-3}. The image in the upper left corner shows the interferogram from which the plasma density was obtained.

be measured interferometrically and the electron density may be calculated using Abel-inversion assuming cylindrical symmetry [95, 96].

Fig. 3.6 shows a plasma density profile at an early time in the channel formation. Detectable ionization by the rising edge of the pulse is represented as an increase in electron density to 3×10^{19} cm^{-3} within about 40 μm in front of the actual channel. The front tip of the channel itself was measured to possess walls at a density of $n_{e,\text{wall}} = 6 \times 10^{19}$ cm^{-3}. This corresponds very well to the plasma density at the channel start as obtained from the comparison of gas density profiles and 2ω-emission from the relativistic channel. The channel is actually expected to have a width of the order of the laser focus or less, because of self-focusing. This lies below the resolution limit of the optical setup. Therefore, the width between the density walls measured by interferometry amounts to $D \simeq 14$ μm. What is actually measured in Fig. 3.6 is the overall plasma density, dropping to zero far from the axis, because there is no ionizing laser present, and having a density depression on axis, which is not fully resolved, but may be attributed to the expulsion of electrons by the laser pulse as discussed in the following section and shown in simulations (Fig. 3.7).

3.4. Simulation of laser plasma interaction

The experiments on electron acceleration presented in this thesis were analyzed using three dimensional particle-in-cell simulations in collaboration with M. Geissler at the Max-Planck-Institut für Quantenoptik in Garching [27, 98]. The simulation code for the laser plasma interaction is called ILLUMINATION [99]. In this code, Maxwell's equations in conjunction with the motion of the plasma's charged particles are solved numerically. The typical size was $144 \times 144 \times 3080$ cells with one particle per cell, corresponding to a simulated volume of $27 \times 27 \times 144$ µm^3 and 64 million particles. To keep the box size small a co-moving window propagating with c is applied to simulate only the laser plasma interaction region. The ions are treated as a immobile neutralizing background. The speed of the simulation is ≈ 3 min/µm on 36 AMD Opteron CPUs.

The code was capable of emulating the experiment very accurately by using experimental parameters for the laser pulse as well as the plasma density profile. The fully ionized plasma to be addressed by the simulation had a density profile as indicated in Fig. 3.5. During the time steps of the simulation the evolution of the electron density, the electromagnetic field of the laser and the energy of the electrons are monitored. Fig. 3.7 shows a sequence of electron density snapshots at different times as the laser pulse is propagating through the plasma. The plasma density in the x-z-plane is shown as colorcode, where x is the laser polarization and z the laser propagation direction. In the first snapshot (a), as the laser enters the simulation volume, the formation of a channel can be seen, where the pulse expels electrons from the propagation axis. The length of this channel corresponds to the laser pulse length. The laser intensity and pulse length are monitored throughout the simulation, but not presented here. As the laser is propagating into higher

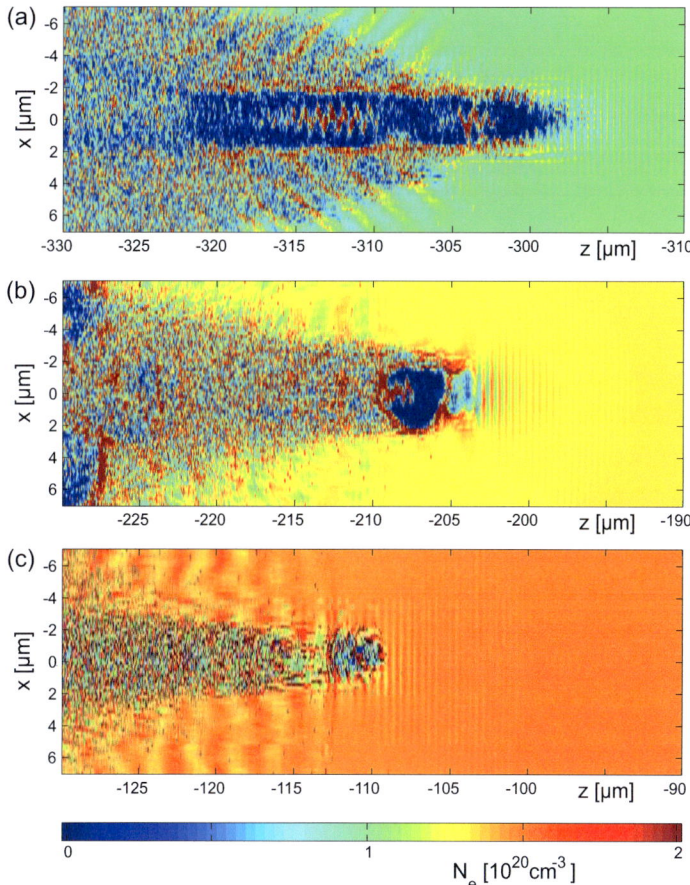

Figure 3.7: Results from 3D-PIC simulation: (a), (b) and (c) show snapshots of the electron density separated by 0.33 ps as the laser pulse is propagating through the plasma in positive z-direction. The electron density in the x-z-plane is displayed as colormap with maximum of 2×10^{20} cm^{-3}. In the sequence (a) to (c) the laser propagation into higher density is modelled according to the measured density profile. First a channel is forming which corresponds to the laser pulse length (a). Propagating into higher density (b) the channel collapses into a single bubble. The bubble consists of an electron void containing a small and very dense electron bunch. In (c) close to the maximum of the density profile the laser's energy is depleted and channel as well as bubble are vanishing.

density (b) the channel shortens as does the pulse length. This snapshot shows also a circular electron void, the bubble, which contains a small and dense electron distribution, the electron stem. Repeated simulations for slightly different parameters always resulted in the formation of a bubble. In (c) where the simulation reaches highest density at $z = 0$ μm the laser is depleted and the channel vanishes. The onset and the end of the channel correspond very well with the experimental observation of the 2ω-emission shown in inset of Fig. 3.5.

3.5. Electron emission

Inside this relativistic channel electrons are accelerated. To monitor the generated electron beam a scintillating screen (Konica KR) was placed 120 mm from the focus in laser propagation direction. It was protected from direct laser irradiation by a 30 μm aluminium filter. The electrons from the relativistic channel have to penetrate the aluminium filter and a plastic carrier layer before inducing light emission from the rare earths (Gd_2O_2S:Tb) in the scintillating layer of the screen. This light emission was imaged onto a synchronized CCD camera.

Aluminium of 30 μm thickness stops electrons with energies below $E \leq 60$ keV and the 300 μm thick plastic layer carrying the scintillating material stops electrons with energies below $E \leq 150$ keV. Electrons of those energies brought to a full stop give only bremsstrahlung at a fraction of $< 10^{-3}$ of their initial energy. This estimation based on data from *Stopping-Power and Range Tables for Electrons* [100], indicates, that the signal observed on the screen is caused by scintillation induced by electrons with energies above ≈ 200 keV only and contribution from bremsstrahlung may be neglected.

From the profile of the electron beam on the screen the beam divergence was measured. The divergence angles given in the following are defined as the full angles of the emission cone. At optimum conditions for relativistic self-focusing, which were determined from the images of plasma self-emission (Fig. 3.5), a very narrow and intense peak of high-energy electrons indicating a beam divergence of

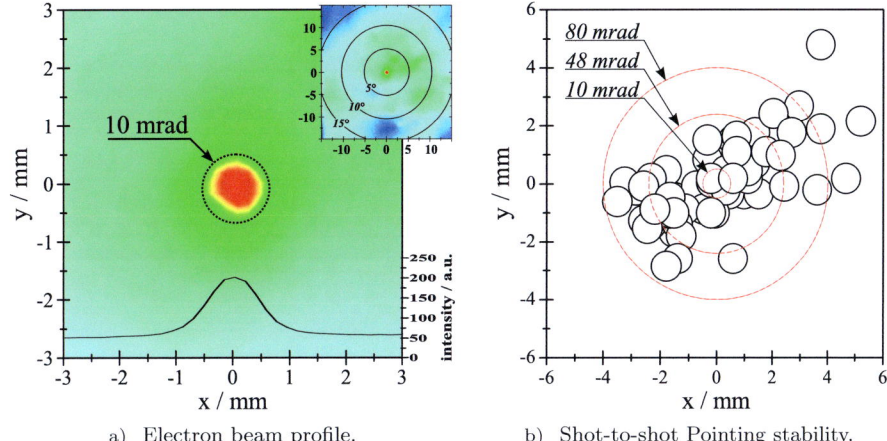

a) Electron beam profile.   b) Shot-to-shot Pointing stability.

Figure 3.8: A scintillating screen was placed 120 mm from the focus and was imaged onto a CCD camera to measure the divergence of the electron beams. (a) The graph displays the intensity distribution on the scintillating screen as well as a line-out along the x-direction at the maximum. The well collimated beam with an full opening angle < 10 mrad sits on top of a pedestal of electrons with a broad divergence of approximately $15°$ (inset). (b) Shot-to-shot pointing stability from 60 consecutive shots. A cone of 48 mrad comprises more than 75% of the electron pulses, while about 93% of the electron pulses were detected within a cone of 80 mrad.

less than 10 mrad was visible (Fig. 3.8a). This peak also exhibited a high shot-to-shot pointing stability (Fig. 3.8b). More than 75% of the electron pulses lay within a cone of 48 mrad and about 93% of the electron pulses were detected within a cone of 80 mrad (determined from 60 sequential shots).

3.5.1. Quasi monoenergetic electron bunches

Recent experiments have demonstrated the capability of laser-plasma accelerators to produce quasi monoenergetic electron bunches [21–23]. Before that achievement laser acceleration of electrons usually resulted in quasi-exponential spectra with one or two associated temperatures. In those experiments the 100 % energy spread proprietary to exponential spectra could be reduced to narrow band spectra. This

Laser system	Laser parameters			Plasma parameters		
	$\tau_L/$fs	$E_L/$J	f/D	$n_e/$cm^{-3}	$\tau_P/$fs	τ_L/τ_P
[23] ASTRA	40	0.5	16.7	2×10^{19}	24	1.7
[21] LOA	33	1.0	18.0	6×10^{18}	45	0.7
[22] L'OASIS	55	0.5	4.0	2×10^{19}	26	2.1
[87] AIST	50	0.1	3.3	2×10^{20}	9	5.6
[88] CRIEPI	70	0.4	5.3	4×10^{18}	57	1.2
[26] JETI	80	0.6	2.2	8×10^{19}	13	6.2

Laser system	Results		
	$E_e/$MeV	N_e	$N_e E_e/E_L$
[23] ASTRA	75 ± 2.3	1×10^8	2×10^{-3}
[21] LOA	170 ± 20	3×10^9	8×10^{-2}
[22] L'OASIS	86 ± 1.7	2×10^9	6×10^{-2}
[87] AIST	7 ± 1.4	3×10^4	3×10^{-7}
[88] CRIEPI	30 ± 6.0	3×10^4	3×10^{-7}
[26] JETI	47 ± 1.5	2×10^6	3×10^{-5}

Table 3.1: Compilation of experiments that observed quasi monoenergetic electron bunches.

constitutes a major advance towards the applications of laser-plasma accelerators for high-energy physics, biology or medicine.

In the above mentioned experiments quasi monoenergetic electron bunches were only observed in a very narrow window of laser and plasma parameters such as laser pulse duration τ_L, plasma density n_e as well as focusing ratio f/D. In particular, the narrow peaks in the spectrum were only generated, if the initial laser pulse duration τ_L matched the plasma period τ_P, analogous to the LWFA condition given in Eq. (2.26). Furthermore, the laser pulse had to be weakly focused into a supersonic gas jet of a few millimeters length at constant density, again owing to the matching of the focus transversely to the plasma period. Typical parameters of these experiments are listed in Table 3.1. If these conditions are fulfilled, the laser pulse drives a so called bubble in the plasma, as explained in section 2.2.3. In the experiment of Geddes *et al.* [22] the necessary conditions were met by self-modulation of the laser pulse in a preformed guiding channel. For the experiments reported in [21–23] this process of bubble acceleration yields narrow peaks in the spectrum at energies E_e of several tens of MeV with energy spreads less than 10 % containing a few 10^9 electrons (electron number N_e). Changing the parameters resulted, again, in an exponential energy spectrum for the electrons without the quasi monoenergetic peaks.

3.5.2. Electron spectra

In contrast to the established scenario, quasi monoenergetic electron bunches were found under considerably different conditions, where the initial laser pulse duration ($\tau_L = 80$ fs) exceeded the plasma period by a factor of 6 (see Table 3.1) [26]. The laser plasma accelerator at high densities was set up according to Fig. 3.3.

The energy spectrum of the electrons was measured by a magnetic spectrometer using an x-ray image plate as detector. The spectrometer was set up and calibrated by B. Hidding from the University at Düsseldorf [26, 101]. The electron spectrometer makes use of the electron deflection in a magnetic field of $B_{max} = 450$ mT. It is able to detect electrons in the range of $\sim 5 \dots 160$ MeV.

The correlation between electron energy and position on the image plate was obtained from three-dimensional numerical calculation of the electron trajectories through the magnetic field. The fringing fields from the magnet and the aperture dimensions ($45 \times 3 \times 5$ mm^3) block electrons with energies < 5 MeV. The spectrometer was placed in z direction 210 mm from the laser focus. The entrance aperture was rectangular and 3×5 mm^2 in dimension. Recalling the pointing stability of the electron beam accelerated from the channel (Fig. 3.8b), between 70 and 90 % of the highly collimated electron beams may be detected, while the others will miss the aperture.

The detector inside the spectrometer is an image plate, which is sensible not only to electrons but also to other particles, x-rays and γ-radiation. Therefore, the image plate was placed off axis, such that it had no direct line of sight to the laser plasma. Furthermore, the housing of the spectrometer was shielded with low and high Z material in order to insure, that only electrons were detected. The response of the image plates is linear to electrons above $1\,\mathrm{MeV}$ with an $1/\cos\vartheta$ dependency, if ϑ is the angle of incidence of the electrons on the image plate [102].

Some of the spectra which result from the image plate analysis are shown in Fig. 3.9. The background may be approximated by averaging the spectra. Between 10 and 30 MeV one finds a quasi-exponential distribution with an average temperature of $T_e = (9.2\pm0.4)\,\mathrm{MeV}$. On top of this broad background peaked features can be seen at energies between 10 and 50 MeV. The last graph in Fig. 3.9 singles out the spectrum containing the most prominent peak. This peak corresponds to a quasi monoenergetic electron bunch with $E_e = (47.0 \pm 1.5)\,\mathrm{MeV}$ containing 2×10^6 electrons.

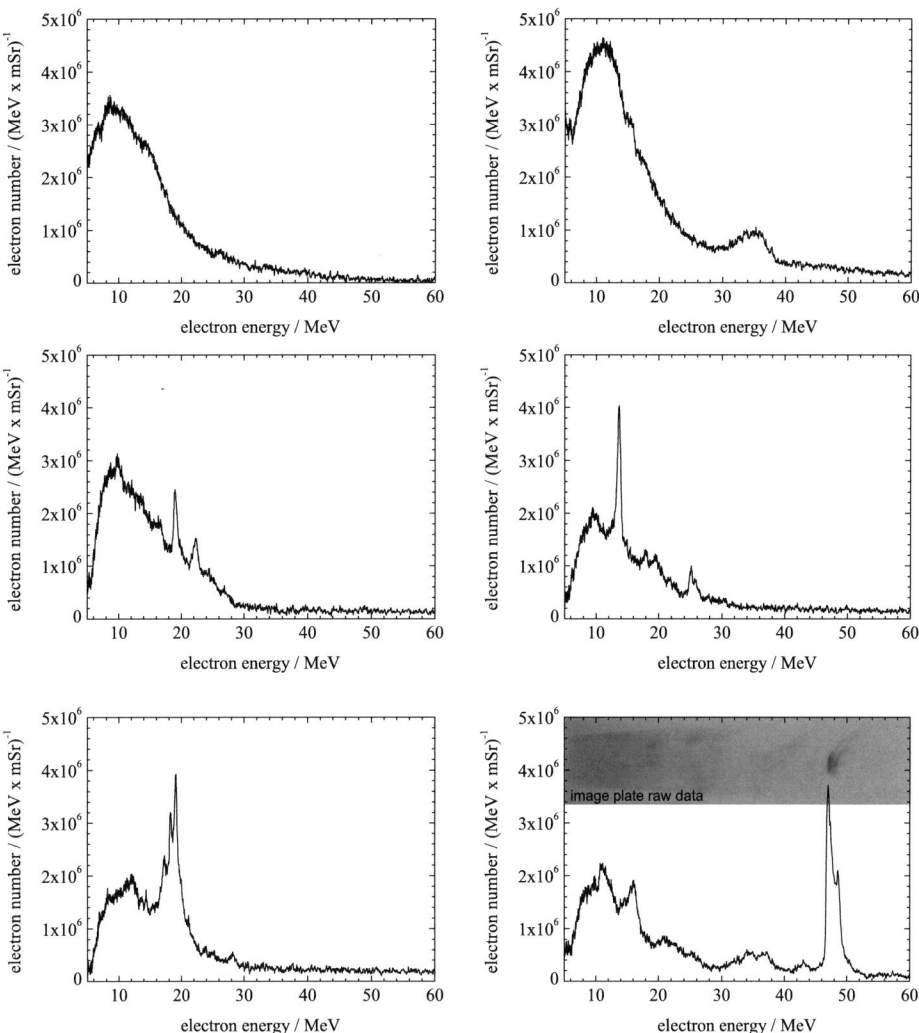

Figure 3.9: A few of the spectra obtained from image plate analysis is displayed in the right graph. All spectra exhibit a quasi- exponential shape between 10 and 30 MeV with an average temperature of $T_e = (9.2 \pm 0.4)$ MeV. On top of this background peaked features appear for some shots. The peak energies lie in the range of $14 \ldots 47$ MeV. The spectrum containing the most prominent quasi monoenergetic peak is shown in last graph. For this spectrum the inset displays the raw data on the image plate. The peak at 47 MeV has a FWHM of ≈ 3 MeV and contains $\approx 2 \times 10^6$ electrons. For this peak the energy is also exceeding the energy of the background distribution.

Self-modulated bubble acceleration

The interaction of the laser pulse with the plasma leading to quasi monoenergetic electron bunches under the given experimental parameters can be, again, understood with help of numerical simulations, as introduced in section 3.4. Input parameters for the 3D-PIC code were chosen to match the parameters for the experiment given in Table 3.1 using a Gaussian density profile. The result for the time evolution of the laser intensity is displayed in Fig. 3.10. The simulation shows, that the initially long laser pulse undergoes strong self-modulation during its propagation in the plasma. The wakefield is generated and leads to longitudinal bunching of the laser pulse envelope as described in Sec. 2.2. The leading part of the pulse is isolated, and becomes shorter. Because the pulse duration is decreased, the condition to drive a bubble as known from the classical bubble regime are fulfilled for the modified pulses.

One may call the mechanism at work here, self-modulated bubble acceleration. The experimentally observed electron bunch energies can be related to laser pulse power requirements for achieving bubble acceleration using the semi-analytical scaling laws discussed in section 2.2.3. Combining Equations (2.29) and (2.30) for a laser at wavelength $\lambda = 0.8\,\mu m$ one can draw a map for the pulse durations and energies needed for bubble acceleration of electron bunches with energies in the range of $E_{mono} \sim 0 \ldots 50\,MeV$, which is displayed in Fig. 3.11(a). The number of electrons expected from Eq.(2.31) is shown in Fig. 3.11(b). For example, the power of a 10 fs pulse with 50 mJ is just slightly above the critical power for bubble generation, and a bunch containing 2×10^9 electrons can be accelerated to about 30 MeV with such a pulse.

On this "map" in Fig. 3.11 one can pinpoint the energies and pulse durations which may overcome the power threshold for bubble acceleration and may lead to monoenergetic electron bunches at energies of $10 \ldots 50\,MeV$. There are two reasons for the percentage of shots resulting in a monoenergetic spectrum.

1. The pointing stability of the electron beam will cause about $10 \ldots 30\,\%$ of the shots to miss the entrance aperture of the spectrometer.

2. The self-modulation of the laser pulse is a highly unstable process. Fluctuations will lead to different energies and duration of the modulated pulse

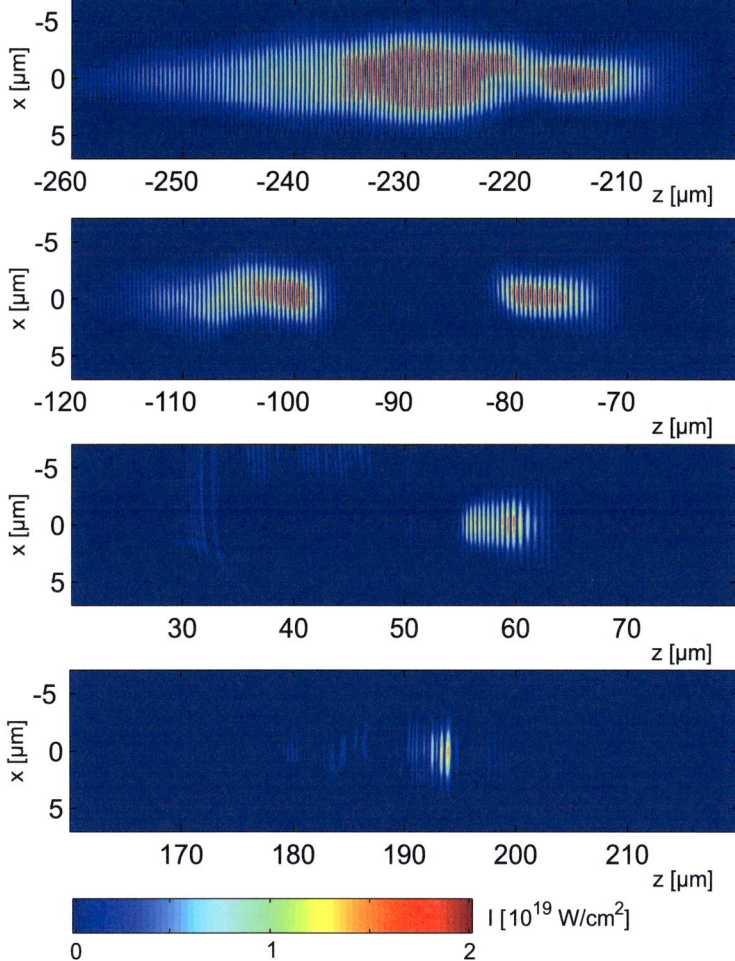

Figure 3.10: Self-modulation and beam break-up. The absolute value of the Poynting vector of the laser pulse is displayed as density plot in the laser polarization plane. The snapshots are separated 45 fs in time, each.

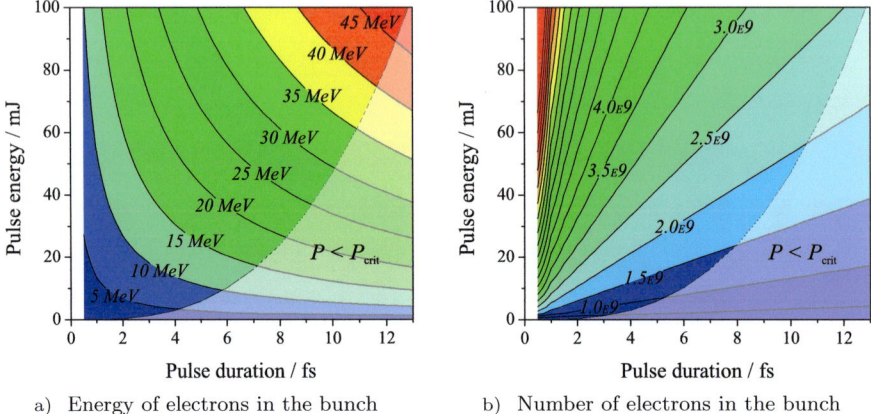

a) Energy of electrons in the bunch

b) Number of electrons in the bunch

Figure 3.11: Scaling laws for monoenergetic electron acceleration. The energy of the monoenergetic peak (a) and the number of electrons in the peak (b) are given as a function of laser pulse duration and energy, based on Equations (2.30) and (2.31), respectively. In the shaded area the laser pulses do not achieve the critical power for monoenergetic electron acceleration (Eq. 2.29). Duration and energy are chosen to reflect the pulses after self-modulation.

fragment, which will drive the bubble. Hence, pulse powers exceeding the critical power are not ensured for every shot and some pulses will not have the ability to achieve critical power while others do.

The simulations for the experiments actually show the self-modulation of the laser pulses to fragments of durations of a few femtoseconds containing 5 . . . 10 % of the initial pulse energy matching the scaling requirements for accelerating monoenergetic bunches to gains observed in the experiment. These scaling laws have been derived for the interaction of a ultrashort laser pulse with a homogeneous plasma under the conditions of Eq. (2.28). The first condition $k_p R \approx \sqrt{a_0}$ is fulfilled for the presented experiment in the density range of $4 \ldots 9 \times 10^{19}$ cm^{-3}, but as the plasma density changes, k_p changes, whereas $R \approx w_0$ and a_0 stay almost constant. Therefore, this scaling statement can complement the physical picture of mechanisms at work in the presented laser plasma accelerator only to a limited extend. The modulated pulses are able to drive a bubble where the energies given by Eq. (2.30) are attained, whereas the number of electron accelerated in this bunch is several orders of magnitude lower than given in Eq. (2.30). The plasma density in experiment presented here was not homogeneous – it had a Gaussian profile, which restricts the time and amount of electrons to be loaded into the bubble.

Furthermore, at the high density the strong focusing is essential to provide good transverse matching conditions of the laser focus to the plasma period. In other words, the transverse bubble radius has to be of the order of half the plasma wavelength, in order to form a stable bubble. This bubble radius can be estimated by equating the electric field of such a bubble with the ponderomotive potential of the laser pulse [60]. From this ansatz one obtains an equation for the initial bubble radius

$$R_0 \approx w_0 \sqrt{\ln \frac{a_0 \lambda_{\mathrm{p}}(z)}{w_0 \pi}} \ , \qquad (3.3)$$

which will arise by focusing a gaussian laser pulse to a waist of w_0 into a plasma at plasma wavelength $\lambda_{\mathrm{p}}(z)$. Analyzing Eq. (3.3) reveals that the bubble radius cannot be perfectly matched to half the plasma wavelength over the whole plasma density profile in the gas jet. But a figure of merit for this matching can be obtained by the relative mismatch between R_0 and $\lambda_{\mathrm{p}}(z)/2$ given by $(1 - 2R_0/\lambda_{\mathrm{p}})$, which is displayed in Fig. 3.12 for a beam waist of $w_0 = 2.5\ \mu\mathrm{m}$. The relative mismatch between bubble diameter and plasma wavelength is minimized in the red valley, which requires rather high laser amplitudes. But for $a_0 \geq 3$ the mismatch between R_0 and $\lambda_{\mathrm{p}}(z)/2$ is less than $20\,\%$ starting from electron density $n_{\mathrm{e}} \geq 5 \times 10^{19}\ \mathrm{W/cm^2}$. This indicates, that the laser power and the focusing in the presented experiment allow for fulfillment of the transverse matching condition and for the generation of a stable bubble near the maximum of the gaussian density distribution.

The simulation of the laser plasma interaction has been performed for several gaussian density profiles changing the peak density. The formation of a bubble could be observed in the simulation for the density profiles with peak densities

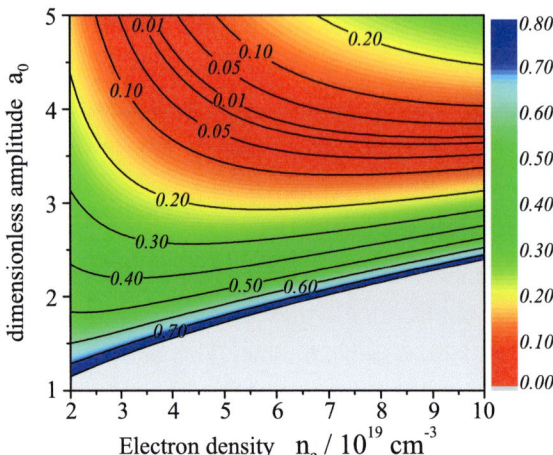

Figure 3.12: Transverse matching of the bubble radius to the plasma wavelength for a beam waist of $w_0 = 2.5\,\mu m$. For $a_0 \geq 3$ the mismatch between R_0 and $\lambda_p(z)/2$ is less than $20\,\%$ starting from electron density $n_e \geq 5 \times 10^{19}\,W/cm^2$.

between $n_{e,min} = 2 \times 10^{19}\,cm^{-3}$ and $n_{e,max} = 1 \times 10^{20}\,cm^{-3}$. One has to keep in mind, that the pulse has to be modulated before it is able to drive a bubble. This process is dependent on plasma density, and therefore the reason for the observed density range for self-modulated bubble acceleration. If the density profile has a peak density below $n_{e,min}$, the self-modulation process is not strong enough and too slow to provide a modified pulse suitable for driving the bubble. On the other hand, if $n_e > n_{e,max}$, the laser pulse reaches the necessary density range for bubble formation too early and the pulse is still too long.

As the particle-in-cell simulation obtains information on the particles as they are exposed to the electromagnetic fields of the laser and the plasma, the simulation itself can be used to monitor the energy spectrum of the electrons. In Fig. 3.13 the electron density distribution at the time corresponding to the last snapshot in the intensity evolution in Fig. 3.10 is displayed. At $z = 193\,\mu m$ the laser is expelling electron that stream around the bubble and will feed the electron stem inside the bubble further. This electron stem constitutes a quasi monoenergetic peak

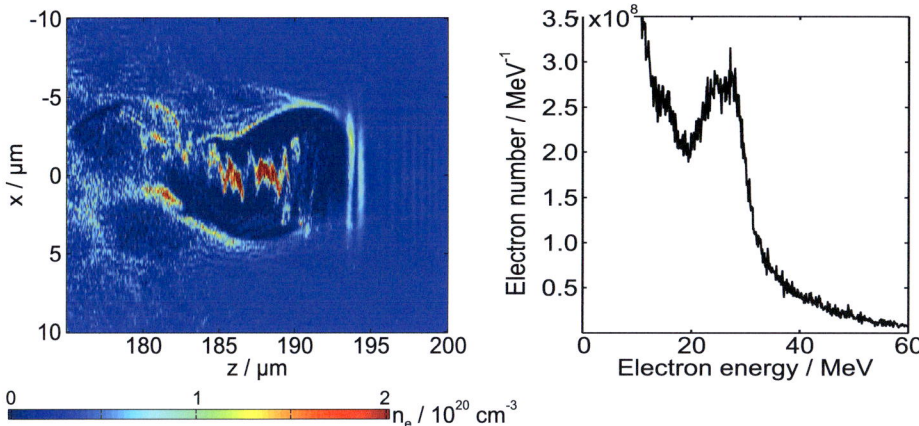

Figure 3.13: The left graph displays the electron density at the time the modulated laser pulse matches the conditions to drive a bubble. On the right an energy histogram of the electrons within the simulation volume is presented. The peak at 25 MeV consists of the electrons in the narrow and dense electron stem inside the while the exponential background has its origin in the surrounding electrons

at about 25 MeV, as presented in the corresponding spectrum in the right graph. The energies for quasi-monoenergetic electrons measured in the experiments are in good agreement with the energies observed in the simulations. The number of electrons in the simulated spectrum is considerably higher than in the experimentally obtained spectrum. The spectrum contains all electrons in the simulation volume at the moment of time when a stable bubble is existent in the plasma. All electron trajectories would have to be analyzed in order to estimate the fraction of electrons that actually reach a spectrometer several centimeters away from the laser plasma.

The discussion of the electron spectra and experiments on laser electron acceleration mentioned above all represent a measurement far away from the actual laser plasma interaction both in space and time. But as the acceleration takes place inside the relativistic channel over a few hundreds of micrometer and femtoseconds it poses a appealing challenge to investigate the acceleration process where and when it actually happens. Time-resolved spectroscopy of the electron acceleration was accomplished in an experiment utilizing the "photon collider" setup at the JETI, where a second laser beam scatters off the laser accelerated electrons in head on geometry. This Thomson-backscattering yields x-rays which were detected and and gave time-resolved information on the electron spectrum inside the channel. The "photon collider" and associated experiments are discussed in great detail elsewhere [18, 20]. The time-resolved optical investigation of bubble acceleration will be the subject of the upcoming sections.

3.6. Tracking plasma bubbles

A short probe pulse was split from the main laser pulse using the transmission of a mirror inside the vacuum beam transport after pulse compression. The transmitted light traversed a telescope to obtain a beam diameter of about 5 mm. The next steps were the second harmonic generation using a 500 µm thick BBO crystal and the beam transport of the probe pulse via a delay line consisting of mirrors mounted onto a µm translation stage.

The frequency doubled probe pulse was deployed as a backlighter via the same optical path and imaging as during the measurement of the relativistic channels at the second harmonic (cf Fig. 3.3 and Sec. 3.3) [103]. This setup enabled the time-resolved observation of the laser plasma interaction by shadowgraphy [96]. The probe pulse was kept at low intensity, such that its propagation in the plasma remains linear. As the time-resolution was delivered by an ultrafast exposure from the short probe pulse, the self-emission from the relativistic channel had to be blocked. This was accomplished by a polarizer, because the polarization of the probe pulse was set to be perpendicular to the polarization of the nonlinear Thomson scattering.

3.6.1. Shadow images

The scheme applied here is different from conventional shadowgraphy, where no imaging is introduced in the optical path. Here a lens is used to image the plasma onto a CCD to have spatial information about the interaction region. The plasma density n_e determines the refractive index η of a plasma $\eta = \sqrt{1 - n_e/n_c}$ with the critical density $n_c = 6.9 \times 10^{21}$ cm^{-3} for $\lambda_{probe} = 400$ nm. The probe pulse is deflected at gradients in the plasma electron distribution which then correspond to dark regions in the shadow images. Representative shadow images are shown in Fig. 3.14. The images have been taken at the same position in space. The main laser pulse is incident from the left. The temporal separation of about 100 fs between each two of the images was accomplished in a multishot experiment with variable delay between main and probe pulse. The time delay was adjusted with an optical delay line mounted on a μm translation stage. The zero delay (0.0 ps) was chosen as the delay, where circular fringes appear in the field of view. The field of view is illuminated homogenously by a probe pulse for the time of less than 100 fs, and in the sequential shots a system of circular fringes can be observed appearing within 0.1 ps, moving along with the laser pulse and disappearing again within 0.1 ps. The source of those ring structures will be discussed in the following section.

Analyzing the position of the ionization fronts and the center of the fringes in the shadowimages for different time delays, one may obtain the velocity of the structure by a linear fit of the time-dependent positions. This was done for irradiation at different intensities (Fig. 3.15). It was observed, that for higher intensities the velocity decreases to about $v_{front} = 0.7\,c$, with c being the speed of light, whereas it is close to c for lower intensities. The reason for this slower propagation of the plasma structure is not clear. As the ionization front of the laser also propagated with this decreased velocity, it is assumed that the ionization process plays a major role for that phenomenon. Therefore, it cannot be observed in the PIC-simulations at the current stage, because ionization was not included.

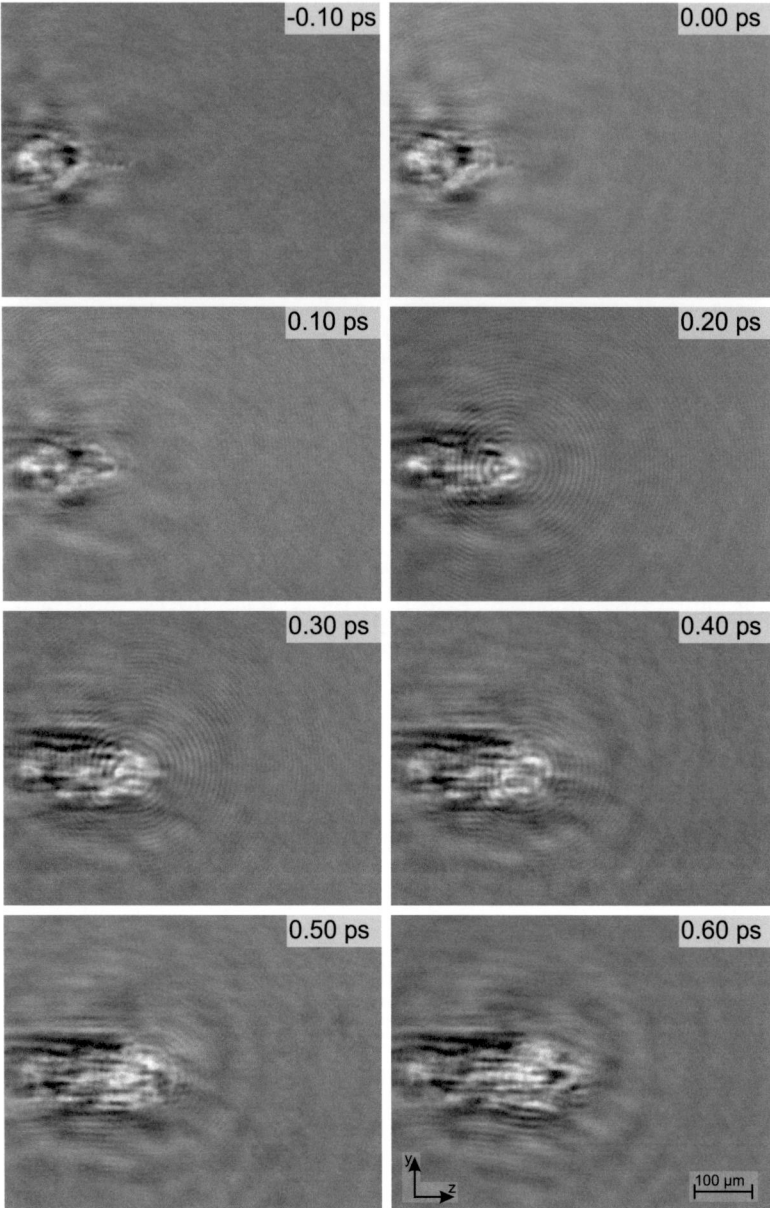

Figure 3.14: Time-resolved shadow images separated in time by 100 fs.

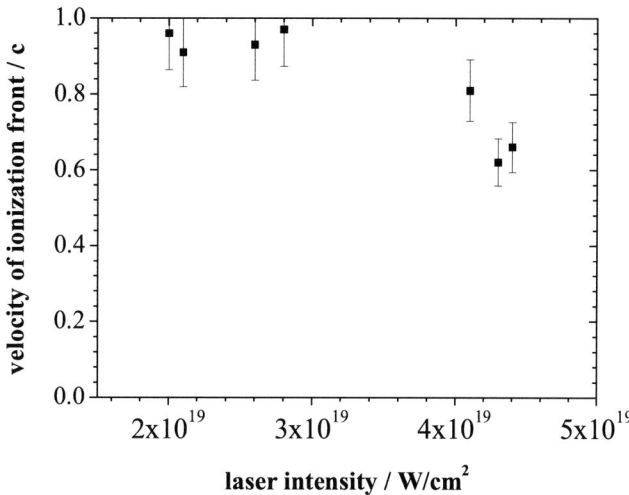

Figure 3.15: Subluminal velocity for ionization fronts and bubbles. At higher intensities the plasma structure is observed to propagate at about $v_{\mathrm{front}} = 0.7\,c$. The velocities for different laser intensities were obtained from the measurement of the position of the ionization front and the center of the fringes in the shadowimages separated by 100 fs. A linear fit of the time-dependent positions then gives the velocity. The accuracy of this measurement amounts to 10% as given from the linear fit.

3.6.2. Ring Structures

Let us pick the fourth picture from the sequence in Fig. 3.14 and take a closer look at it (Fig. 3.16). At the time this image was recorded the main laser pulse, incident from the left, had already propagated about 220 µm into the field of view. Thus, it has passed the position of its vacuum focus and travelled about two thirds of its self-focused distance. It has ionized the He gas along its path. The relativistic channel has been formed, but its emission is blocked by a polarizer.

The shadow image also exhibits fringes, visible over the entire field of view, even at positions were neither main laser pulse nor plasma are present. This is a very reproducible feature in a wide range of main laser intensities. The object causing the fringes is smaller than the resolving power of the optical setup $(5 \ldots 10\,\mu m)$, and thus cannot be observed directly. The center of the fringes coincides with the position of the laser pulse. The fringes are almost circular and the distance between successive fringes decreases with increasing radius. When the time delay between the main laser pulse and the backlighting probe pulse is changed the interference pattern can be seen moving along with the main laser pulse. This movement can be observed in the sequence Fig. 3.14.

In the experiment this interference effect can only be observed in a very short time interval of $\Delta t \approx 0.5$ ps while the main laser pulse is propagating in its relativistic channel. There are no fringes for earlier times or later times. The fringes were observed for each shot for those probe pulse delays, which indicates that this phenomenon is caused by a very stable structure.

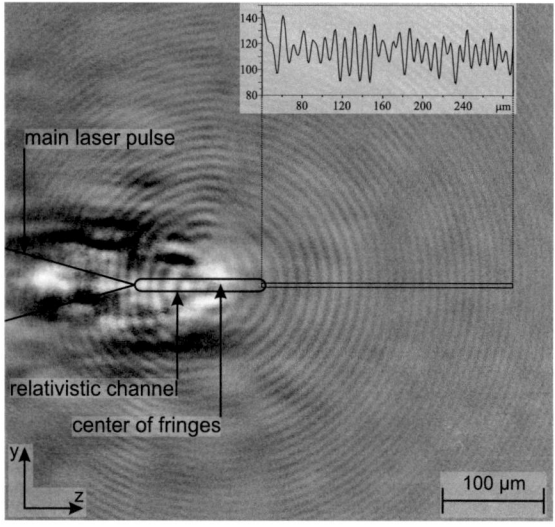

Figure 3.16: Shadow image of the laser gas interaction. The laser pulse is incident from the left and ionizes the gas along its path causing intensity modulations in the shadow image. Furthermore fringes, which are almost circular, are visible over the entire field of view. At the time when this image was recorded the main pulse has propagated to the position tantamount to the center of the fringes. The area where the relativistic channel resides is indicated in the image. Self-emission from the channel is blocked by a polarizer.

3.6.3. Raytracing for shadowimages

The ring structure, described in section 3.6.2 may be interpreted to be caused by the deflection of the central part of the probe pulse by a small and dense plasma peak. The main part (vacuum bundle) of the probe pulse will take no notice of the small plasma peak and will reach the CCD unperturbed while the central part (plasma bundle) is deflected. Thus, the small plasma acts as a defocusing lens or even point-like source for the deflected part of the bundle. Interference of the deflected part with the part that remains unperturbed leads to the fringes that are visible in the shadow images.

In order to discuss the ring structures, the deflection, imaging and interference of the probe pulse were simulated with a two dimensional phase sensitive raytracing algorithm (Fig. 3.17a) [27]. Light paths were calculated through a given plasma density distribution, represented by its index of refraction $\eta(\vec{r})$, using the ray equation

$$\frac{\mathrm{d}}{\mathrm{d}s}\left(\eta\frac{\mathrm{d}\vec{r}}{\mathrm{d}s}\right) = \operatorname{grad}\eta(\vec{r}), \tag{3.4}$$

where $\vec{r}$ is the position vector of a point on the ray and s is the length on the ray [104]. In order to consider interference effects, it is necessary to keep track of the optical path and thus of the phase of each individual ray. Furthermore, it is essential to take into account spherical aberration of the real lens for the imaging (Fig. 3.18b), since the part of the rays that are deflected by the plasma is not paraxial anymore. Therefore, this part does not form an image of the bubble on the CCD, but will be again diverging when it reaches the image plane. The wavefronts of the plasma bundle and the vacuum bundle will interfere in the image plane, yielding circular fringes.

The graphs Fig. 3.18 show the result from a simulation. In the left graph, the black curve shows the retrieved intensity on the CCD if only the amplitude of the individual rays is detected without taking the optical path into account. If the phase of the rays is considered the superposition of the rays gives the interference fringes (blue curve). The right graph (Fig. 3.18b) shows the result from two

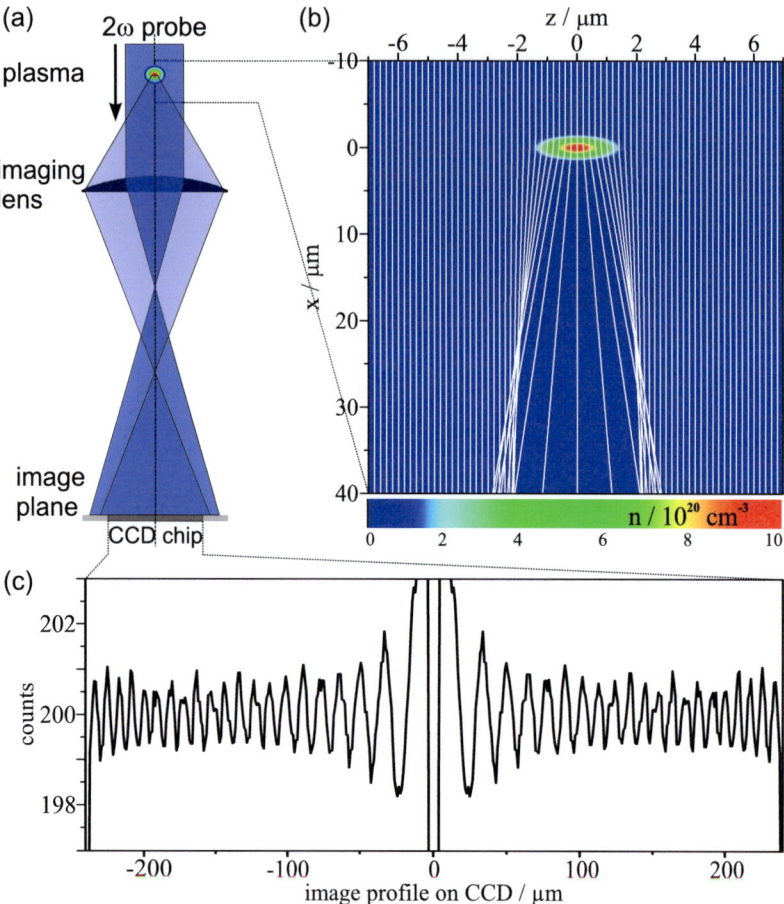

Figure 3.17: Raytracing: (a) shows a sketch of the situation addressed by the raytracing algorithm. The probe pulse is illuminating the plasma which is imaged by a lens onto a CCD camera. In (b) the region around the plasma at the origin is magnified. It shows the propagation of a bundle of initially parallel rays contained in the probe pulse propagating in positive x direction. A small and dense plasma with a Gaussian distribution is located at the origin and models the electron stem inside the bubble. Rays are deflected according to the change of refractive index induced by the plasma. In (c) the raytracing result on the CCD chip is displayed. The interference of the deflected part of the bundle and the part that is not affected by the plasma appears as fringes on the detector.

phase sensitive simulations. It clearly shows the necessity of simulating the real curvatures of the imaging lens. Using an ideal lens in the raytracing only a noisy but constant distribution on the CCD is observed, whereas the simulation with the real lens and correct curvatures gives rise to the fringes. The raytracing results which are discussed for the influence of the plasma were all obtained under the consideration of the phase of the rays, as well as the imaging parameters of the real lens.

In Fig. 3.17b some rays are shown as they are propagating in positive x direction. The Gaussian shaped plasma distribution at the origin represents the small and dense electrons located inside the bubble from Fig. 3.19 which is deflecting the central part of the ray bundle. Imaging the plane of $x = 0$ for an assumed Gaussian electron distribution with peak density of 9×10^{20} cm^{-3} and full width at half maximum of 1.7 μm leads to an interference pattern in the image plane, which extends over the whole detector, (Fig. 3.17c) similar to the fringes observed in the experiment (Fig. 3.16).

The raytracing yields interference fringes for an assumed Gaussian electron distribution, whose peak density and width is in correspondence to the values for the electron stem from the bubble observed in the PIC simulation within a factor of

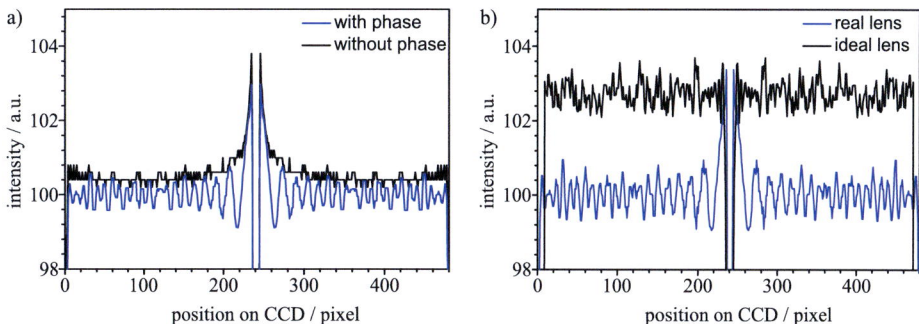

Figure 3.18: Comparison of raytracing with and without taking the phase information of the individual rays into account (a). Also the imaging characteristics of the real lens are essential for the reconstruction of interference fringes in the simulation.

2. The frequencies of the interference patterns from simulation and experiment are in good agreement. The separation between adjacent fringes decreases with increasing distance from the center. The contrast of the fringes observed in the experiment is an order of magnitude larger than from the simulation.

The raytracing results depend on two parameters, the peak density and the width of the Gaussian. If the peak density is reduced, the extension of the in-

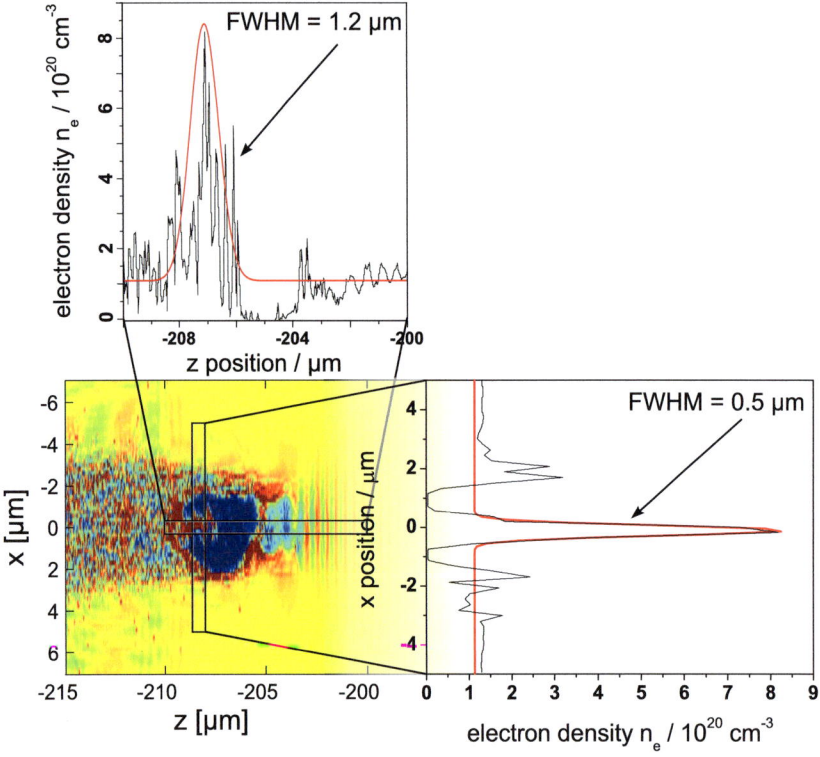

Figure 3.19: Density profile of a bubble. The line-out from the electron density distribution obtained from the PIC-simulation shows that the density of the electron stem inside the bubble exceeds the density of the bubble walls and the ambient plasma by a factor of eight and is the prominent source for the deflection of the probe beam. The longitudinal and lateral line-outs show different profiles.

terference pattern on the detector is decreasing, i. e. a high density is needed to generate fringes over the whole field of view. The contrast of the fringes increases with increasing width of the Gaussian distribution, i. e. if more rays are deflected, the contribution to the interference is higher. Matching the contrast from the simulation to the values obtained from the shadowgrams may only be achieved for Gaussians that would be wide enough to observe them directly.

The simplifying assumption of a Gaussian profile with spherical symmetry was derived from the density profiles of the bubble (Fig. 3.19), because the electron stem is the prominent feature in the density profile. The bubble walls attain twice the background density, whereas the electron stem exceeds the ambient plasma by a factor of eight. The simulation was also performed with the more realistic bubble profile comprising the bubble wall and the evacuated region. It was found in the raytracing, that those components do not contribute significantly to the interference fringes, because their density and extend are considerably smaller than the ones of the electron stem.

The relativistic channel also affects the propagation of the probe pulse, and might therefore play a role in the image formation. However, the densities reached in the channel walls are expected to be similar to the ones of the bubble wall, i. e. less than the density achievable in the electron stem. Also the fact that the circular system of fringes appears within 0.1 ps and is then moving as a whole with the laser pulse is an argument in favor of the electron stem. The interference pattern caused by a channel would be elongated, and is not expected to move on the time scale of 0.1 ps. Even for high plasma temperatures of the order of 1 MeV, the expansion velocities of channels are in the range of a few percent of the speed of light [105], leading to changes of a few μm within a time span of 1 ps.

The interference pattern was observed in the experiment for the time of the laser plasma interaction only. The interpretation is corroborated by reconstruction of the fringes in a raytracing simulation for the electron stem. Therefore, the fringes may be attributed to the dense electron bunch inside the bubble, and thus, represent the first direct experimental evidence of a bubble plasma structure.

§ 4 Proton Acceleration

In this chapter the laser-based accelerations of protons and ions will be discussed. Table-top laser systems similar to JETI have demonstrated the generation of proton beams with energies that would be suitable for inducing (p,n) reactions in nuclei [106]. Therefore, it was desirable to characterize the potential of the JETI laser system for proton acceleration. The logical path for this pursuit was to reproduce and learn from results that have been obtained with other laser systems (see Sec. 2.3 and literature given therein). The influence of target material and laser parameters on the proton/ion-spectra had to be identified. Therefore, some of the results of proton acceleration experiments at the JETI will be presented, which give an insight in the capabilities for proton generation at the JETI. The next step was the experimental investigation of the concept of proton acceleration from microstructured targets, which lead for the first time to laser-acceleration of quasi-monoenergetic protons [29].

4.1. Experimental setup

For the acceleration of protons one employs a different accelerator setup than for the electrons and moves from underdense plasmas to overdense plasmas. The laser will be directed onto a thin foil which enables TNSA to generate fast proton beams (cf. Sec. 2.3). The schematic setup is on display in Fig. 4.1. The JETI laser pulses are again focused by an f/2.2 off axis parabolic mirror. In the focal spot of typically 5 μm^2 size intensities of 4×10^{19} W/cm^2 are reached.

Thin metal foils were placed in the focus as target. Protons and ions accelerated from the foil were detected either by a beam imaging system (Colutron Inc.), comprising a chevron microchannel plate (MCP) with phosphor screen and CCD camera, or on nuclear track detector plastics CR39. Before hitting the detector, the proton/ion beam traversed a Thomson parabola type spectrometer comprising a

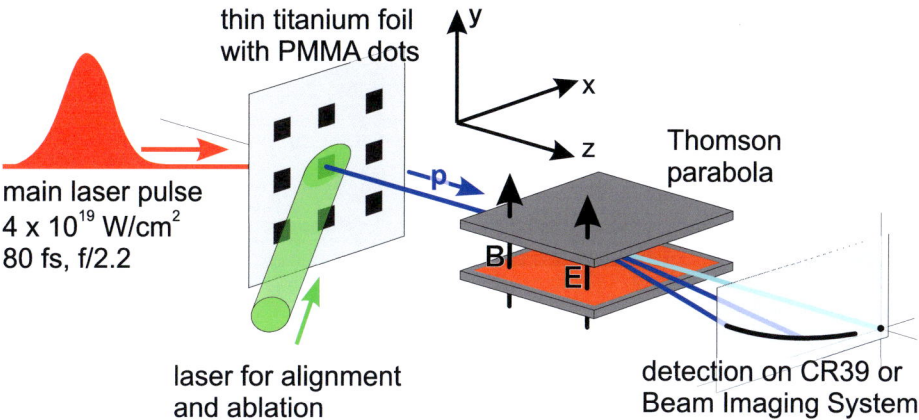

Figure 4.1: Experimental setup for proton acceleration. The laser was focused onto a thin metal foil. Target details will be discussed in the text. The protons and ions accelerated from the target are dispersed with respect to energy and charge-mass ratio in a Thomson parabola and then detected by either a beam imaging system, comprising a micro-channel plate (MCP) with phosphor screen and CCD camera, or on nuclear track detector plastics (CR39).

magnetic field ($B = 535\,\mathrm{mT}$) parallel to an electric field ($E_{\mathrm{max}} = 2500$ V/cm) [107]. Thus, the proton/ion beam was dispersed with respect to energy and charge-mass ratio. The energy distribution was then determined from the spatial distribution of the protons on the detector.

The nuclear track detector plastics is one of the most widely used detection methods for proton and ion tracks. CR39 is a polymer that is sensitive to high energy particles, such as protons, ion and neutrons. The energy deposition of the particle inside the CR39 causes damage in the polymeric bonds. Consecutive etching of the material for two hours in a 6M NaOH solution at a temperature of $85\,^{\circ}\mathrm{C}$ leads to local destruction of the CR39 at the sites where it was struck by a particle [108]. This etching results in pits in the CR39, that are associated to one particle each and can then be counted under a microscope. Therefore, the ion tracks on CR39 yield an absolute measurement of the ion spectra.

The downside of using CR39 is the long and weary process from the laser shot to the spectrum by etching and counting the CR39. This also goes along with the

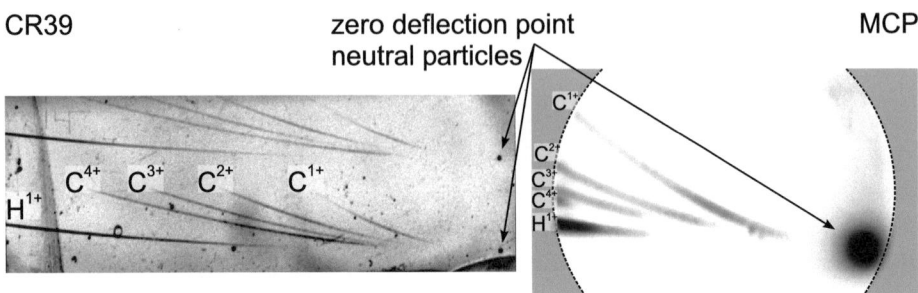

Figure 4.2: Ion tracks detected with CR39 and MCP. Neutral particles will not deflected by the fields and gives a calibration point for the spectra. The horizontal distance from this point is directly related to the particle energy. The left image shows two sets of ion tracks on a CR39 plate. The MCP on the right gives tracks of lower quality and in a restricted energy range, but allows for on-line detection.

fact, that with this setup only three spectra can be recorded before the vacuum interaction chamber has to be let up again to take out the irradiated plastics. In order to take advantage of the high repetition rate of the laser system an on-line diagnostic for the proton spectra had to be achieved. For that purpose a beam imaging system was deployed. It consists of two chevron microchannel plates at a voltage of 1 kV each. The MCP represents an array of tubes for the generation of secondary electrons from impinging particles. The particles give rise to secondary electrons, that are accelerated by the applied voltage. Since the tubes are tilted with respect to the acceleration direction, the secondary electron hit the walls of the microchannel multiple time, and induce an electron avalanche. With this process an amplification of 10^4 may be attained. The electrons emitted from the microchannel plates traverse a voltage of 3 kV before they hit a phosphor screen and induce light emission from the screen. This light is imaged via a fiber taper and a camera objective onto a CCD camera. The MCP system therefore allows for spatially resolving measurements of the ion beam. Therefore, the Thomson parabolas may be detected.

The MCP system has the major advantage of providing real time detection, whereas it does not give an absolute signal for the detected ions. Therefore, it was carefully calibrated by a comparison of pixel counts to proton pits on a piece of CR39. The CR39 plastics were irradiated at the same laser and target conditions as the comparative shots using the MCP system. The evaluation of the respective spectra yielded a minimum MCP sensitivity of (6 ± 1) protons/count [107]. The MCP system has a higher threshold for proton detection than the CR39. Therefore, detection via MCP gives good data concerning yield, temperature and shape of the proton spectra, whereas it cannot reach the dynamic range of the CR39 detection. The MCP system is nevertheless the detection of choice, owing to its high flexibility and real time observation of the spectra.

The energy spectra from MCP and CR39 are measured with an energy resolution of 50 keV, which corresponds to the resolution of the MCP, and per solid angle of 10 µSr defined by the 3 mm aperture in front of the Thomson parabola.

4.2. Protons from plain targets

The thin foils used as targets in the presented experiments comprised Titanium and Tantalum, and are commercially available from Goodfellow Inc. They were stretched in a rigid aluminium frame to achieve optimum flatness, such that the target can be moved to an intact area for each new laser shot without changing the surface position with respect to the laser focus [89]. With motorized translation stages inside the target chamber the target can therefore be moved in the x-y-plane to enable several hundred shots on the same target without breaking the vacuum. Utilizing a high repetition rate laser, such as JETI, together with the MCP on-line diagnostic enables a fast and systematic exploration of proton acceleration from thin foils. During an experiment many shots may be recorded, which allows for flexible investigation for different parameters with high reproducibility.

Optimum thicknesses and materials

For the given laser parameters the optimum target conditions had to be identified. Therefore, Ti and Ta foils of different thicknesses were deployed. The laser pulses exhibit amplified spontaneous emission that is preceding the high-intensity part of the pulse. If no further measures are taken to suppress the ASE level, it has an energy contrast of 10^{-6} and a duration of $\tau_{\mathrm{ASE}} = 5 \ldots 6$ ns (cf. Sec. 3.1, [94]). Fig. 4.3a shows proton spectra from titanium foils at different thicknesses, that were recorded for this ASE level. Higher proton yield and increased cutoff energy were obtained when decreasing the thickness of the irradiated foils. This enhancement in proton yield and energy continued until an optimum thickness was reached (Fig. 4.3a, black curve: Ti at 2 µm). For titanium at 1 µm the proton acceleration was less efficient. Kaluza *et al.* [73] found that the optimum thickness is determined by the ASE duration. The ASE prepulse is inducing a shock wave that is travelling through the foil. Thus, a density gradient builds at the target back side before the TNSA mechanism becomes effective. If the scale length of

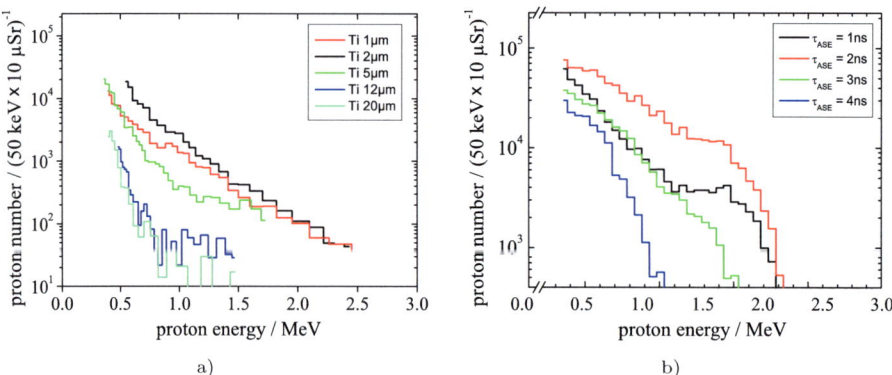

a)   b)

Figure 4.3: Optimum thickness and ASE. (a) The spectra from titanium foils at different thicknesses. The fast pockels cell was not deployed ($\tau_{\mathrm{ASE}} = 5 \ldots 6$ ns). (b) Effect of ASE suppression using the fast pockels cell. Titanium 2 µm. Please note, that the spectra in (a) were detected on CR39, while (b) resulted from MCP detection.

the back side plasma exceeds the Debye length the proton energy will decrease.

In the succeeding experiment the laser system was equipped with a supplementary pockels cell between regenerative amplifier and 4-pass amplifier. This pockels cell with a fast rise time of less than 0.5 ps provided an additional suppression of the ASE and the prepulse in the laser system. An intensity contrast between ASE and main laser pulse of the order of $10^{-9} \ldots 10^{-10}$ was attained (cf. Sec. 3.1, [94]). The duration of the ASE prepulse can be adjusted by changing the time delay of the pockels cell gate with respect to the Ti:Sa pulse. When the pockels cell timing was adjusted to provide a short time span of ASE in the range of $\tau_{ASE} = 0.5 \ldots 5$ ns preceding the main pulse, an optimum duration for a 2 μm thick titanium foil can be found (Fig. 4.3b) [73]. The shortest ASE duration will in principle not be the optimum duration, because the ASE provides the preplasma for the electron acceleration and plasma heating.

4.3. Protons from coated and microstructured targets

In order to enhance the proton yield in comparison to plain targets, the foils have been coated with polymethylmetacryalate [PMMA, $(C_5O_2H_8)_n$] on the back side. So called double layer targets have been investigated to give an increase in proton energy and number [109, 110]. Titanium and Tantalum foils of 2 μm and 5 μm have been coated with a homogeneous PMMA layer of 1 μm and 0.5 μm thickness. In the irradiation with laser pulses, this lead to an enhancement of the proton yield by factors of 5 to 15 as compared to plain foils (Fig. 4.4).

As the Titanium foils showed better performance both in proton yield and maximum energy as well as in handling for the experiment, they were the target of choice for the experiments on proton acceleration from microstructured targets in the prospect of producing monoenergetic protons, following the theoretical proposition of [78] argued in Sec. 2.3. Therefore, the coated foils were subjected to femtosecond laser ablation in order to generate the micro structures (dots) on the target back side. This process yielded dots with a size of (20×20) μm^2 with a

PMMA "free" space of 50 μm in between (Fig. 4.5). The extension of the dots differs slightly owing to the fabrication process.

Alignment of the dots in the laser focus

In the experiments each laser pulse had to hit the target on the front side exactly opposite to one of the dots on the target back side. A frequency doubled Nd:YAG laser (532 nm) was illuminating the target back side in order to localize the dots (Fig. 4.1). The PMMA was doped with Rhodamine 6G and when the dot was moved into the focus of the alignment laser, the yellow fluorescence light (around 600 nm) indicated, that the dot was brought into position of the focal spot on the front side for irradiation. The foci of the accelerating pulse and the alignment laser were brought to spatial overlap. An imaging system consisting of an objective lens generating an intermediate image, that is observed via a microscope objective with a CCD camera, was used as a visual control of the target back side [111]. The position of the foil in the Ti:Sa focus was determined from the γ-dose measured

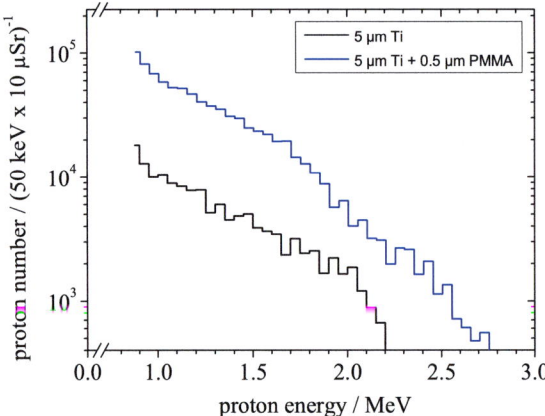

Figure 4.4: Enhanced proton yield from double layer targets. Applying a polymer layer on the back surface of the foil allows for enhanced proton yield from the laser acceleration. The graph above shows that the proton number from a 5 μm thick Ti-foil may be increased by a factor of 6 ± 2, if a PMMA layer of 0.5 μm is present on the back side. The spectrum from the plain foil is an average over 4 shots, whereas the one for the coated foil is an average of 8.

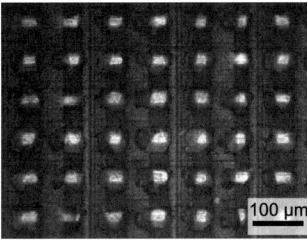

Figure 4.5: Microstructured target for proton acceleration under the microscope. The titanium foil of 5 μm thickness carries dots of PMMA with a thickness of 0.5 μm and a transverse size of (20×20) μm². The PMMA dots on the back side of the Titanium foil were produced by grid-like femtosecond laser ablation of the PMMA layer.

by an ionization chamber placed outside the vacuum chamber. In consequence the focal position of the Ti:Sa on the foil was determined by generating a small hole in the foil with an attenuated Ti:Sa pulse. This position was fixed on the back side observation and the Nd:YAG focus was directed to this site as well.

The laser could also be used to clean the backside from the parasitic proton layer by laser ablation before the high intensity laser struck the target. Heating is expected to reduce the hydrocarbon contaminants on the target back side. Thus, the yield of heavier ions than protons will increase, while the number of protons that are accelerated is diminished [68, 112, 113].

When the laser pulse struck the foil exactly opposite a dot the data obtained with the MCP changed significantly. The left panel of Fig. 4.6 shows MCP images for the two cases of irradiation. The zero deflection points and the proton tracks are visible. For shots on an unstructured and uncoated area of the foil, the proton track becomes more intense with increasing distance from the zero deflection point, whereas the proton track from a dot exhibits its intensity maximum closer to the zero deflection point. The analysis of the proton tracks with respect to intensity and energy yielded the usual exponential spectra for the unstructured area. Exemplary spectra are displayed in the graph on the right of Fig. 4.6. For the acceleration from the unstructured area an average over 6 shots

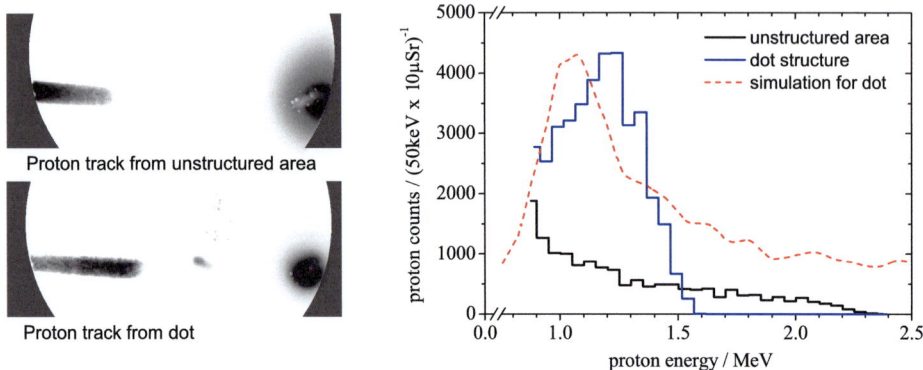

Proton track from unstructured area

Proton track from dot

Figure 4.6: Reduced bandwidth for proton acceleration. On the right two raw images from the MCP show the proton tracks from an unstructured part, and a dot, respectively. The corresponding spectra are displayed in the graph on the right. Laser plasma acceleration from a dot clearly results in a peaked spectrum (blue curve) in excess of a exponential background (black curve, average over 6 shots). The red curve is the spectrum from the corresponding simulation.

is given. The acceleration from dots resulted in peaked spectra. The laser plasma acceleration from a dot gives rise to a peaked spectrum (blue curve). Here no average is given, because the position of the peaked feature is changing slightly from shot to shot. The narrow band spectra exhibit peaks at energies of about $E = 1.3 \pm 0.2$ MeV. Also the number of protons in the peak fluctuates in the range of $10^4 \ldots 10^5$ protons/10 µSr from shot to shot, because the dots on the back side are not uniform in size. The quasi-monoenergetic proton bunch in Fig. 4.6 contains about 4×10^4 protons/10 µSr. The varying dot size gives also rise to the fluctuating full width at half maximum of the peaks in the range of $\Delta E = 200 \ldots 600$ keV. However, a peaked spectrum was generated with very high reproducibility, when the foil was irradiated opposite to a dot. The red curve in Fig. 4.6 shows the spectrum from a simulation for the laser and target parameters for the shots. The simulations confirm the experimental observation very accurately, and will be described in detail in Sec. 4.4.

The peaked spectra were also observed using CR39 detection. The CR39 was used in order to give spectra with higher dynamical range than the MCP, and to

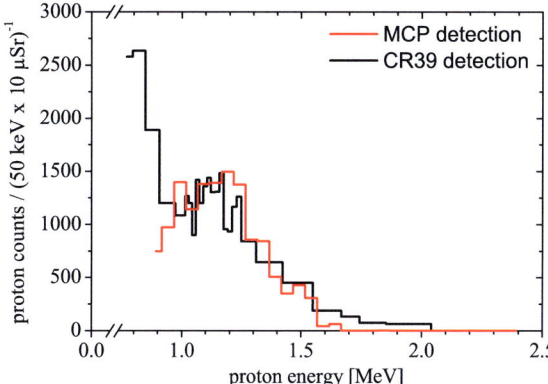

Figure 4.7: Peaked spectrum detected with MCP and CR39. The spectrum from an irradiated dot detected on the CR39 exhibits a proton peak in agreement the measurement using the MCP. The single proton counting on CR39 allows for a measurement wit higher dynamical and spectral range. Therefore, the exponential background from parasitic protons is more pronounced.

extend the measurable spectral range to lower energies. The single proton counting on the CR39 also reveals a peaked spectrum in agreement with the spectra observed with the MCP system. An exponential background for lower energies was observed on the CR39. These parasitic protons originate from the contamination layer on the target. There will also be a low energy contribution from the dots that lie next to the one which was actually irradiated. The dots were separated by 50 µm, whereas the proton source size may be larger than that. This fact shows, that the heating of the target back side using the Nd:YAG laser as described above will have to be optimized for further reduction of parasitic protons.

The spectra depend crucially on the target design, and the appearance of the narrow band feature in the spectra from irradiated dots is due to the enhanced proton yield from the dot, that is situated within the center of the TNSA field. An intuitive rationale for the generation of narrow-band spectra can be obtained from the potential of a pancake shaped electron bunch situated in front of a thin foil (Fig. 4.8). The accelerating field at the foil at $z = 0$ may be calculated for an

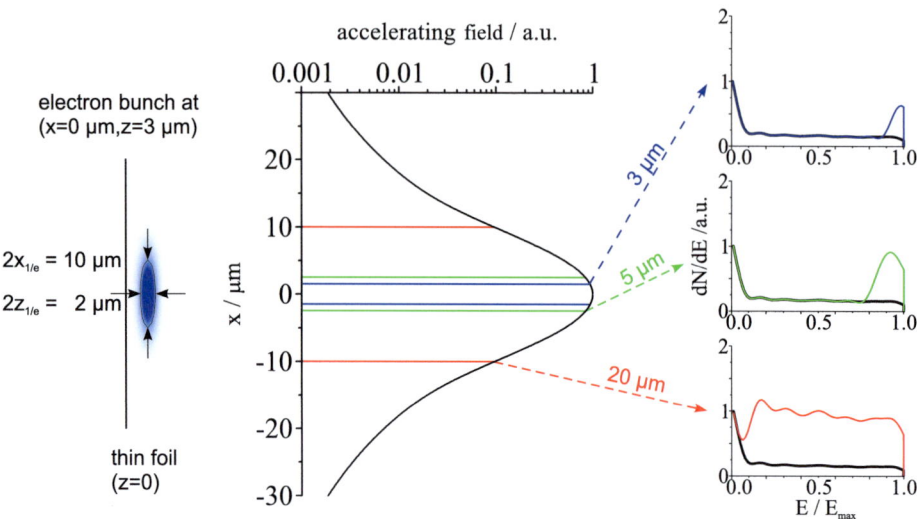

Figure 4.8: Sampling the field of an electron bunch. The geometry of the problem is shown in the left panel: An electron bunch with gaussian density distribution is situated in front of a conducting surface. The field at this surface is displayed in the middle. Dots in the center of the field of 3 μm, 5 μm, and 20 μm width give increased proton yields corresponding to the range that is sampled by the dot.

electron density in the bunch, assumed to be gaussian in both directions,

$$\rho_e(x, z) \sim \exp\left[-\frac{x^2}{x_{1/e}{}^2}\right] \exp\left[-\frac{z^2}{z_{1/e}{}^2}\right]. \tag{4.1}$$

The full 1/e-width is $2\,z_{1/e} = 2$ μm in longitudinal direction and $2\,x_{1/e} = 10$ μm in transverse direction. This simplifying model gives the correlation between the accelerating field and the spectra that are obtained if a confined range of the field is sampled (Fig. 4.8). The black curves in the right hand side graphs correspond to spectra from an homogeneous contamination layer accelerated in depicted field. If the proton layer is increased by a factor of five in the center of the field, i. e. if a dot is applied, then the proton yield is enhanced in the energy range that is sampled by the dot. This yields peaks in the spectra, whose bandwidth depends on the extension of the dot in comparison to the one of the field (blue, green and red curves in the right hand side graphs of Fig. 4.8).

The value for the energy spread of the obtained spectra $\Delta E/E \approx 25\,\%$ is not yet low enough to call the protons monoenergetic. The dots on the target back side were still considerably larger than the laser focus, $400\,\mu m^2$ vs. $5\,\mu m^2$. The extension of the proton rich area is thus sampling a much larger area than the nearly homogeneous range of the accelerating field, and the energy spread of the peak is increased. However, the dot size is still well below the overall source size, resulting in a peaked spectrum. Although Fig. 4.8 is based on a rather simplified assessment of the situation, it implies an interesting approach for future experiments, namely the measurement of the accelerating field at the target back side. With dots smaller than the laser focus and a positioning of the dots relative to the focus with a accuracy of the same order, it would be possible to sample and measure the transverse shape and extend of the field.

In order to get a more reliable indication for the relation between laser/target parameters and the spectra observed from the dots, one has to resort to simulations that address the complexity of the process.

4.4. Simulation and scalability

The analysis of the presented experiments is supported by two-dimensional particle-in-cell (2D-PIC) simulations performed by Timur Esirkepov from JAERI [114], based on the code REMP, the model described in [115], and the given experimental parameters . Fig. 4.9a shows the geometry addressed in those simulations: The laser is incident on the thin Titanium foil under $45°$, where first the interaction of the ASE and the rising edge of the pulse is treated hydrodynamically (HD) to infer pre-plasma conditions. The left hand side of Fig. 4.9a shows a simulation snapshot after 50 laser cycles. The two-dimensional HD simulation initializes the conditions for the subsequent 2D-PIC simulation of plasma expansion, electron acceleration and transport, and consequently the proton/ion acceleration from the target back side. In the right hand side of Fig. 4.9a at a simulation time of 250 laser cycles the protons and carbon ions may be observed as they detach from the foil and propagate in normal direction.

The code used in the simulations is highly parallel and was run on the 720-processor HP AlphaServer at JAERI. The PIC simulation tracked 6.3×10^9 cells and 1.4×10^{10} quasi-particles in total. For the simulation the laser parameters pulse energy, pulse duration, focus size, as well as ASE duration and contrast are considered. Target parameters include foil thickness and dot size. A contamination layer is not present in the simulation. Fig. 4.9b and c plots the resulting

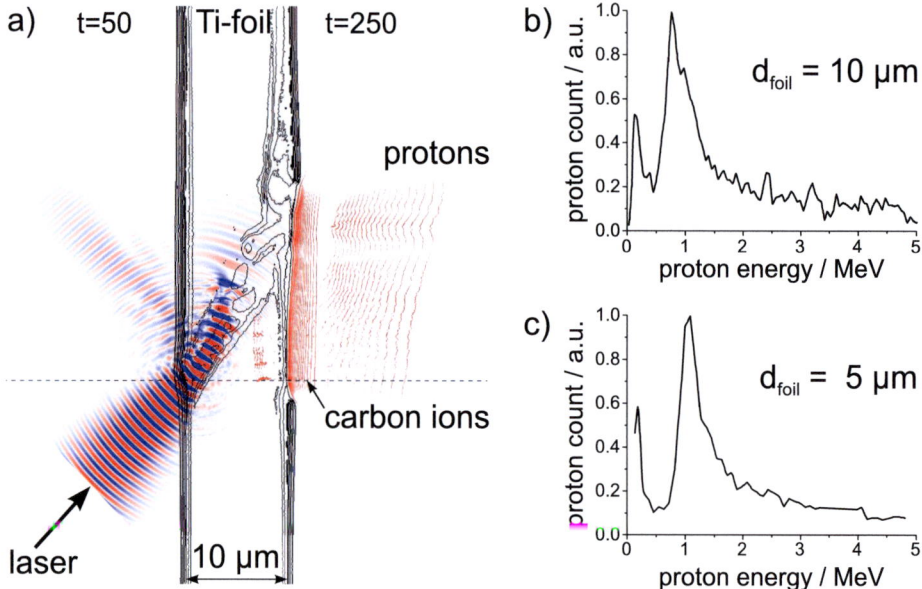

Figure 4.9: 2D-PIC Simulation for Dot-TNSA by T. Esirkepov. a) The laser pulse hits the thin Ti-foil, where its interaction and electron acceleration is modelled hydrodynamically. Panel a) consists of two simulation snapshot, the left hand side shows the laser and the density of the foil at $t = 50$ laser cycles. Whereas the subsequent plasma expansion and proton/carbon acceleration from the PMMA dot is simulated with a PIC code, and displayed for $t = 50$ laser cycles on the right hand side. Panels b) and c) show the resulting spectra for different thicknesses of the titanium foil.

proton spectra (solid line) under the conditions $I = 3 \times 10^{19}$ W/cm^2, and 20 µm
PMMA structure on the back side of a 10 µm and 5 µm thin titanium target, respectively. The proton spectra obtained from the simulation exhibit a narrow band
structure around $E = 1.2$ MeV with a full width at half maximum $\Delta E = 0.3$ MeV
or $\Delta E/E = 25\%$. The simulated spectrum for the titanium foil of 5 µm thickness
is also plotted in the graph of Fig. 4.6. The comparison of the curves from simulation and experiment clearly shows the good agreement with respect to both the
existence of the narrow structure as well as its position and width. The simulation shows, that the protons are emitted within a cone comprising a solid angle
of 24 mSr. With the numbers obtained from the experiment lying between 10^4
and 10^5 protons/10 µSr, the acceleration of a quasi-monoenergetic bunch containing about 10^8 protons in total may be expected. In contrast to the experimental
results, the spectra obtained from the simulation do not exhibit an exponential
background, because the foil in the simulation is equipped with a "pure" dot, and
does not contain parasitic protons.

It was estimated in [32, 78], that in the case of a 10 µm focal spot and 0.1 µm
thick proton layer, a proton energy spread of 1% can be expected. Furthermore,
from simulations [32, 77] it is known that for micro-structured targets the maximum proton energy scales as the square root of the laser power with

$$E_{\text{max}} = 230 \,\text{MeV} \times \left(\frac{P}{1 \,\text{PW}} \right)^{1/2} . \qquad (4.2)$$

This is a convenient scaling, but it is restricted to laser pulses at intensities above
10^{20} W/cm^2, clean of ASE and prepulses. Eq. (4.2) also requires an optimum for
a parameter, called critical depth $\sigma = n_e \, d$, which represents the product of the
electron density of the target n_e and the target thickness d. The optimum critical
depth also scales with intensity $\sigma_{\text{opt}} \approx n_{\text{cr}} \lambda \, (3 + 0.4 \, a_0)$. For a given target material
($n_e = const.$) the optimum target thickness determines the optimum energy ab-
sorption. If the target is too thin, laser energy is transmitted through the foil, while
for thicker foils a considerable amount of laser energy will be reflected instead of

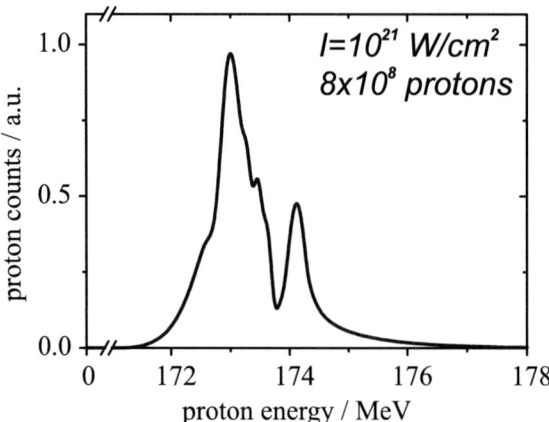

Figure 4.10: Protons expected from *Polaris*. The simulation of proton acceleration from a 5 μm thick titanium foil with 0.1 μm thick PMMA dots of 2.5 μm diameter. The laser parameters have been chosen to reflect the specifications of the upcoming *Polaris*: laser pulses of 150 J within 150 fs focused to an intensity of $I = 10^{21}$ W/cm^2. The simulation then yields a narrow bandwidth peak at 173 MeV.

contributing to the acceleration. In the case of titanium ($n_e = 5.68 \times 10^{22}$ cm^{-3}) the optimum target thickness for a laser at wavelength of about 1 μm amounts to 0.27 μm.

The numerical tools at hand, it is worthwhile to extrapolate the technique of proton acceleration from microstructured targets to future experiments. The laser development at the *Institut für Optik und Quantenelektronik* is concentrated on the construction of a high repetition rate table-top laser system with Petawatt power (*Polaris*). This laser system uses all-diode-pumped Yb^{3+}-doped fluoride phosphate glass at a center wavelength of 1032 nm and will reach the Petawatt level within the next years [116].

Proton acceleration using microstructured target is one of the first experiments coming to mind for the scientific case of the *Polaris*-system. In order to estimate the outcome of such an experiment, the parameters for the simulation have been changed to smaller dot sizes and higher energies. *Polaris* will deliver pulses of

150 J within 150 fs which leads to an intensity of about $I = 10^{21}$ W/cm^2 in focus of 100 µm^2. The dot on the back side of a 5 µm thick titanium foil was assumed to have a diameter of 2.5 µm and a thickness of 0.1 µm. The simulation for these laser and target parameters resulted in a proton peak energy at 173 MeV and an energy spread of $\Delta E/E \sim 1\%$. Under those conditions, the proton yield is not longer limited by laser energy, but all protons contained in the dot (8×10^8) are accelerated to an energy of (173 ± 1) MeV.

Since *Polaris* was assumed to deliver a power of 1 PW to the target, the question arises, how the discrepancy between the scaling given in Eq. (4.2) and the simulation result of 173 MeV comes about. The scaling assumes a laser pulse free of any prepulses and ASE, while the *Polaris* simulation included an ASE of contrast 10^{-7}. Furthermore, as titanium at 5 µm is used in the simulation, the critical depth as discussed above is not at its optimum.

§ 5   Conclusion and Applications

The implementation of laser plasma accelerators for electrons and protons, as well as the experiments and its discussion presented above, represent an important contribution to the research on laser-based particle acceleration. It was demonstrated, that laser plasma accelerators are in principle capable of generating highly collimated, high-energy, monoenergetic electron and proton beams. The demonstration that monoenergetic particle beams from laser plasma acceleration will be possible, opens the door to an increased variety of applications.

Some aspects of potential and future application for laser plasma based accelerators will be discussed in this chapter, such as injectors for contemporary accelerators, as high-energy particle and light sources at reasonable size and cost for university-scale research and in medical cancer therapy.

5.1. A future option for particle acceleration

In order to implement a laser plasma accelerator as an injector for contemporary accelerators, the generated particle beams have to reach high energies with low energy spread [117, 118]. Concepts that may lead to the achievement of this requirement have been presented above. Apart from the energy requirements on the particle beam another important parameter is the emittance of the beam. A low emittance means good coherence or directionality. The beam coherence determines, for instance, the maximum beam propagation distance, the force that is needed to maintain good beam transport, the minimum focal spot size, and applicability as an injector to subsequent acceleration stages [119].

The normalized transverse emittance of a particle beam is given by $\varepsilon_N = \beta\gamma\sigma_x\sigma_{x'}$, where β and γ are energy parameters of the beam, and σ_x and $\sigma_{x'}$ represent the source size and the beam divergence, respectively [28]. The experiments presented here gave quasi-monoenergetic electron bunches gaining energies up to $\gamma \sim 100$, $(\beta \approx 1)$. They were emitted with a divergence of less than 10 mrad. Assuming a source size of about 10 μm, the normalized transverse emittance may be estimated to be $\varepsilon_N < 10$ mm mrad. For the narrow band proton beams the source size was 20 μm with a divergence of about 170 mrad. Due to the sub-relativistic energies of $E = 1.3$ MeV for the proton beam the relativistic factor attains $\gamma \approx 1.001$ and $\beta \approx 3 \times 10^{-3}$. The normalized transverse emittance therefore amounts to $\varepsilon_N \approx 10^{-2}$ mm mrad. In the case of the beam expected from *Polaris*, the parameters $E = 173$ MeV, $\sigma_x \approx 2.5$ μm and $\sigma_{x'} \approx 17$ mrad also allow for an normalized transverse emittance of $\varepsilon_N \approx 10^{-2}$ mm mrad, because for higher γ a reduced divergence is expected.

For the longitudinal emittance, being the product of the energy spread and the bunch length $\varepsilon_{\text{long}} \approx \Delta E \, \Delta\tau$, also minimum values represent optimum beam characteristics. Owing to the fact, that laser plasma accelerators operate on a time scale of several hundred femtoseconds, the bunch lengths are expected to be of the same order. The electron bunch length may be inferred from electro-optical sampling using THz emission from the electron bunch as it passes the plasma-vacuum boundary and was found to be of the order of $\Delta\tau \sim 50$ fs [120–122]. For the electron accelerator presented here, an upper limit for the electron bunch length of $\Delta\tau < 300$ fs was obtained from such THz-emission measurements [123]. Therefore, the longitudinal emittance is expected to achieve values as low as $\varepsilon_{\text{long}} < 10^{-6}$ eV s for the presented electron bunches ($\Delta E \approx 3$ MeV). For the proton bunches from microstructured target a duration $\Delta\tau < 10$ ps may be assumed, and the reduced energy spread of $\Delta E \approx 0.5$ MeV also implies a longitudinal emittance in the order of $\varepsilon_{\text{long}} \sim 10^{-6}$ eV s.

In current accelerator technology the values for normalized transverse emittance and longitudinal emittance lie in the order of $\varepsilon_N \sim 1$ mm mrad and $\varepsilon_{\text{long}} \sim 0.5$ eV s, respectively [28, 117]. The laser generated electron and proton beams reach values well below the ones from conventional accelerators, and thus, laser plasma accelerators inherit a higher performance with respect to those parameter.

Particle bunches in conventional accelerators contain between 10^7 and 10^{10} particles. Therefore, the particle number in the bunches from the laser plasma accelerator for electrons ($\sim 10^6$ for JETI and $\sim 10^9$ for other groups, see Tab. 3.1), as well as the one for protons ($\sim 10^8$) also match the requirements for application.

In the prospect of achieving higher energies research has to concentrate on gaining control over the plasma processes in order to exploit the high gradients to full extend. In the case case of proton acceleration the approach of sampling the accelerating fields was demonstrated here. The future of laser plasma accelerators for electrons lies in the elongation of the acceleration length using preformed channels and plasma cells. Another challenging technique to further increase the electron energy is *staged acceleration* using several consecutive wakefields for acceleration of the electron bunch. This could be demonstrated last year for the first time at the Naval Research Laboratory [124]. Electrons at about 1 MeV from 2 TW laser driven nitrogen plasma were injected in a second laser plasma accelerator driven by a 10 TW pulse in a helium gas jet, and were accelerated to about 20 MeV in the second stage.

5.2. Applications for laser accelerated electrons

Laser accelerated electrons may be applied as high brilliance light sources over the whole electron magnetic spectrum. Terahertz radiation may be generated from laser accelerated electron bunches [125]. Sending the electron bunches through bending magnets or undulators in a FEL setup may also yield high-energy photon [121, 126]. In an all optical setup the backscattering of light from the relativistic electron promises an versatile x-ray source most suitable for pump-probe experiments due to their short duration, high brightness and inherent synchronization (cf. [18, 20] and references therein).

Apart from the copious amount of potential application of laser accelerated electrons, in the following an example will be discussed, how electrons may be immediately used to generate high-energy bremsstrahlung, leading to table-top laser nuclear physics.

Laser nuclear reactions and isotope production

The energies of several tens of MeV carried by laser accelerated electrons may be converted to γ-rays, i.e. bremsstrahlung, by the interaction of the electron beams with high-Z material. The efficiency for this conversion lies in the order of 10% [17]. The γ-rays achieve energies of the order of 10 MeV, which matches the giant dipole resonances (GDR) of nuclei. Therefore, the irradiation of elements may lead to the excitation of those GDRs and consequently to fission of the nucleus or emission of neutrons or protons from the nucleus. The concept of *laser nuclear physics* was identified by Boyer et al. in 1988 [127], ten years before the first experimental demonstration of laser induced (γ, n)-reactions and neutron production from laser induced nuclear fusion at the NOVA laser [75]. Laser nuclear physics has been a very active field of research [10–13, 89, 128–130] using solid target for both electron acceleration and subsequent bremsstrahlung conversion. In the following, it will be discussed, that the separation of the two steps, i. e. accelerating the electrons from a gas target and using a high Z material for bremsstrahlung conversion, resulted in an increased efficiency [17], because electrons from gas jets reach higher energies than those from the small-scale preplasma in front of a solid target.

The photo-nuclear reactions are also very useful as a diagnostic tool for laser generated electron jets at high energies [131]. Leemans *et al.* demonstrated the feasibility of electrons accelerated from a gas jet with a table-top laser to induce photo-nuclear reactions in copper. Nuclear reactions with thresholds as high as 20 MeV were verified [130].

The photon field generated in such experiments has unique properties related to the way it has been produced: high energy, high flux and short pulse length are achieved. Therefore, it is challenging to obtain information about the bremsstrahlung photon field. Dose measurements based on thermoluminescence detectors [9, 132] are only suitable at low photon fluxes and temperature around 1 MeV. At higher fluxes and temperatures, as obtained from the laser accelerators, the photon field may only be characterized through photo-induced nuclear reactions [13, 133, 134].

77

The whole diversity of laser induced nuclear reactions using solid target has been demonstrated at the JETI [89]. Results obtained for nuclear reactions induced by bremsstrahlung from electrons accelerated from gas jets, will be presented in the following.

It is possible to obtain information about the energies of the photon field from photo-nuclear reaction yields, if two or more reactions can be induced and their cross-section values are known [133, 134]. The photo-nuclear cross-sections in the GDR regime are usually approximated with a modified Lorentzian [135]:

$$\sigma(E) = \sigma_{\max} \frac{(E\Gamma)^2}{\left(E^2 - E_{\max}{}^2\right) + (E\Gamma)^2} \, , \tag{5.1}$$

where Γ is the width, $E_{\max}$ and $\sigma_{\max}$ are position and value of the maximum cross section, respectively. In Fig. 5.1 the cross sections for $^{181}\text{Ta}\,(\gamma, \text{n})\,^{180}\text{Ta}$ and $^{181}\text{Ta}\,(\gamma, 3\text{n})\,^{178}\text{Ta}$ are displayed.

Because of the low cross sections for photo-nuclear experiments, an accumulation of several thousand shots is required to obtain a detectable amount of nuclear reactions. Therefore, it is reasonable to assume a relativistic Maxwellian energy distribution function for the averaged energy spectrum of the laser-accelerated electrons. This assumption has been made in earlier experiments [11] and still holds true in the case of electrons from the gas target, since only a fraction of the shots exhibited monoenergetic features in the spectrum which themselves were not stable in energy. For high energies the energy distribution can be approximated with a Boltzmann function. In the ultra-relativistic case the generated bremsstrahlung photons will approximately have the same temperature as the electrons [136]. In the intermediate case in the range of several MeV, the photon temperature will be slightly lower than the electron temperature [132]. Owing to the threshold values for the GDRs only photons with energies $E > 7\,\text{MeV}$ give a contribution to the photo-nuclear reactions. Thus, those photons are assumed to be Boltzmann distributed:

$$\frac{\mathrm{d}N_\gamma(E)}{\mathrm{d}E} = N_{\gamma 0} \exp\left[-\frac{E}{T_\gamma}\right] \, , \tag{5.2}$$

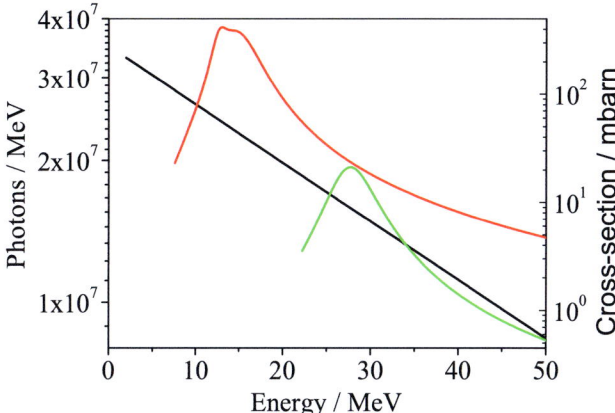

Figure 5.1: Shown are a Boltzmann distributed photon spectrum of a temperature of $T_\gamma = 35\,\mathrm{MeV}$ according to Eq. (5.2) (left scale, straight line) and the photo-nuclear cross-sections of the (γ, n) and $(\gamma, 3\mathrm{n})$ reactions in $^{181}\mathrm{Ta}$ (right scale, red and green lines, respectively), JENDL evaluations from [135]. The integral over the photon spectrum times the cross-section determines the reaction yield (Eq. (5.3)).

with the amplitude $N_{\gamma 0}$ (in units of MeV^{-1}), and the temperature of the spectrum T_γ. In this case the photon temperature represents a lower limit to the temperature of the electrons from which the photon field is produced ($T_\mathrm{e} \geq T_\gamma$). For a given reaction type with cross section $\sigma(E)$ and threshold energy E_th, the number of induced reactions can be calculated by

$$N = \rho_\mathrm{a} d \int_{E_\mathrm{th}}^{\infty} \sigma(E)\, N_\gamma(E)\, \mathrm{d}E\,, \tag{5.3}$$

where ρ_a is the atomic density of the target and d the target thickness. Measuring and comparing the reaction yields of at least two reactions with different cross sections one can unambiguously determine the temperature and the amplitude of the photon spectrum. Fig. 5.1 shows the exponential photon spectrum and the

Isotope	Reaction	E_{th}/MeV	$E_{\text{max}}/\text{MeV}$	$\sigma_{\text{max}}/\text{mbarn}$	FWHM/MeV
^{197}Au	(γ, n)	8.1	13.5	529	4.5
^{197}Au	$(\gamma, 3\text{n})$	23.1	27.1	14	6.0
^{181}Ta	(γ, n)	7.6	12.8 & 14.9	221 & 330	2.1 & 5.2
^{181}Ta	$(\gamma, 3\text{n})$	22.1	27.7	21	5.6

Table 5.1: GDR parameters of the observed photo-nuclear reactions [135]. The (γ, n) cross-section of ^{181}Ta exhibits a double-peak structure because also $(\gamma, \text{p} + \text{n})$-reaction occur in that energy range.

photo-nuclear cross sections of the relevant reactions in ^{181}Ta which determine the reaction yield. The largest errors in this calculation are usually introduced by uncertainties in the photo-nuclear cross section data.

Photo-nuclear reactions in ^{197}Au

In one of the preliminary experiments a 2 mm thick ^{197}Au plate was used as converter material irradiated by 10^4 laser shots. Fig. 5.2 shows the γ-spectrum of the sample integrated over the first 100 h after irradiation. From the time-resolved γ-spectra reaction yields of 2.32×10^8 for the (γ, n) reaction to ^{196}Au as well as 3.23×10^4 $(\gamma, 3\text{n})$ reactions to ^{194}Au could be determined. All lines corresponding to ^{196}Au are marked with a circle. The spectrum in Fig. 5.2 also shows that ^{196}Au was not only produced in the ground state but also in an excited state with an exotic angular momentum of 12$-$ and consequently a comparably long half-life of 9.7 h. In the (γ, xn) reactions so many neutrons were produced that also about 6×10^5 neutron capture reactions in ^{197}Au were detected. The corresponding line is marked with an open square in Fig. 5.2. From the reaction yields of the (γ, n) and $(\gamma, 3\text{n})$ reactions the photon temperature was determined to be 4.5 MeV.

Photonuclear reactions in ^{181}Ta

^{181}Ta was used as converter target in most activation experiments since the products from photo-nuclear reactions are rather short-lived enabling the reuse of the targets in subsequent experiments and were detectable with a setup of time-resolved γ-spectroscopy. As in ^{197}Au the reaction yields of the (γ, n) and $(\gamma, 3\text{n})$

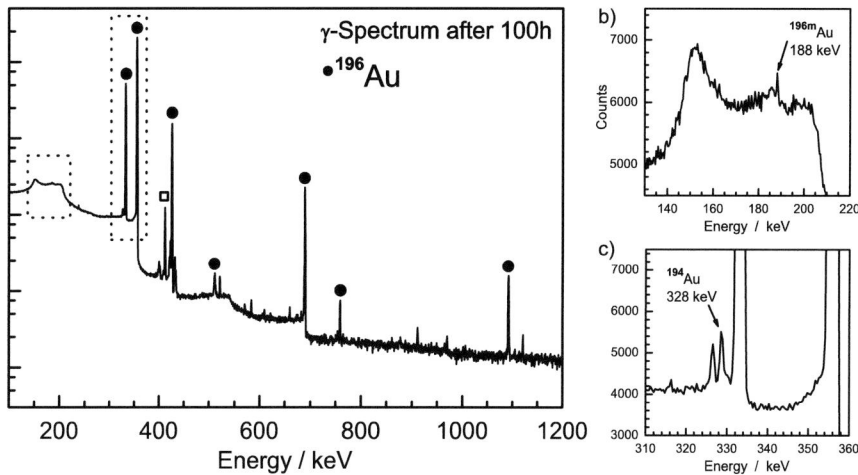

Figure 5.2: a) γ-spectrum of the ^{197}Au converter target integrated over the first 100 h after irradiation. Lines related to ^{196}Au are marked with filled circles. An open square marks the 412 keV line of ^{198}Au The spectrum is displayed in logarithmic scale, the signals observed are well above background. Dotted regions of the spectrum are magnified below. b) 188 keV line of ^{196m}Au. c) 328 keV line of ^{194}Au.

reactions were used to measure the temperature of the photon spectrum averaging over 10^4 laser shots. Separate experiments with slightly varying laser conditions resulted in photon temperatures ranging from 10 to 35 MeV. The large fluctuations of the temperature measurements indicate a very sensitive dependency of the acceleration process on laser parameters.

The photon temperatures determined in these experiment clearly show, that bremsstrahlung generation from electrons accelerated in a gas target is more efficient than irradiating the solid targets directly. Because in the latter case the attainable temperatures are limited due of the ponderomotive scaling law [64].

Transmutation of ^{129}I

A very interesting isotope for exploration of laser nuclear reactions is ^{129}I. It is very long-lived and is a major constituent in waste from nuclear energy production. By laser-induced transmutation it can be converted to the short-lived ^{128}I which will decay to Xenon in about 25 minutes. This transmutation had been demonstrated using a bulk converter target irradiated directly by the laser and its cross section was just measured recently [89, 137]. The experiment was repeated using the gas-jet for electron acceleration as shown above and consecutive bremsstrahlung generation in a converter. With the electron spectra at higher numbers and temperatures the reaction yield for iodine transmutation could be increased by two orders of magnitude as compared to the earlier experiments [17]. An encapsulated sample of PbI$_2$ containing ^{129}I was placed directly behind the primary ^{181}Ta converter target. Time-resolved γ-spectroscopy after irradiation with 10^4 laser shots, showed the characteristic 443 keV line for the decay of ^{128}I to ^{128}Xe and the 443 keV from the beta emission from ^{128}I. Thus, a total yield of 2.16×10^5 reactions ^{129}I $(\gamma, n)\,^{128}$I could be determined.

The photon temperature was determined from the (γ, n) and $(\gamma, 3n)$ reaction yields in ^{181}Ta to be (9.8 ± 1.5) MeV in this experiment. Applying Eq. (5.3) with the measured photon field and reaction yield of iodine transmutations, the cross section of ^{129}I $(\gamma, n)\,^{128}$I may be determined. For this calculation the cross section was assumed to have a shape given by Eq. (5.1) with Γ and E_{max} derived from the stable ^{127}I. Therefore, the maximum reaction cross-section value of the (γ, n) reaction was determined to be $\sigma_{\mathrm{max}} = (250 \pm 100)$ mbarn using a method solely relying on photo-nuclear reactions [17, 89, 137].

5.3. Applications for laser accelerated protons/ions

Apart from serving as an injector for subsequent acceleration, more applications for laser induced proton beams are conceivable. The short proton beams may be used as a probe for high electric fields in laser plasma interactions [138, 139]. Furthermore, proton beams may be of use as a fast ignitor for laser fusion [140]. The ignition relies on the fact, that protons penetrating matter, deposit their energy at the end of their trajectory, leading to efficient heating of the pellet irradiated for fusion. This characteristic energy deposition of protons and ions may also be exploited for the generation of therapeutic particle beams in medicine.

Laser based medical cancer therapy

One intriguing application conceivable for laser accelerated protons and ions is radiotherapy in oncology [141]. In cancer therapy the tumor cells are subjected to radiation in order to diminish or even destroy the tumor. The ionization inside the tissue leads to radicals and subsequent bond breaking in the DNA strands. The ionizing radiation can be delivered by x-rays, γ-rays, protons or heavier ions [142]. The deposition of energy inside the irradiated tissue depends strongly on the type of radiation and the energy that is used.

When x-rays traverse into material they deposit their energy with an exponentially decreasing distribution. In order to reduce the dose rate in the healthy tissue in front of a deep sited tumor the photon energy has to be increased and x-rays are replaced by high-energy γ-rays – first from the radioactive isotope ^{60}Co, which decays to ^{60}Ni under the emission of two γ-rays with energies of 1.17 MeV and 1.33 MeV, and for even higher energies of up to 18 MeV from the bremsstrahlung of electrons accelerated in LINACs. The target volume (tumor) is then irradiated from different directions using intensity modulated beams to achieve a high dose rate in the tumor in a conformal way, whereas attaining a reduced integral dose in the surrounding healthy tissue. Therefore, the major issue for radiation therapy is the high contrast between *high target dose* and *reduced integral dose* [143]. The ability to generate such high contrast target volumes is the essential advantage of proton and ion beams over electromagnetic radiation for radiotherapy. Protons and ions show an inverse depth dose profile. The tissue is ionized at the expense of the energy of the proton until the proton is stopped. At the end of the proton range the energy deposition reaches a very sharp maximum, also known as the Bragg peak. Deep sited tumors with complicated geometry can, therefore, be irradiated with significantly increased dose contrast.

In Fig. 5.3 protons and carbon ions are compared with respect to their stopping and depth dose profiles in water, which may be used as a rough approximation for biological tissue. The stopping power was obtain from tables that can be found in [100, 144]. It determines the energy loss of a particle as it traverses matter, and thus the range that ions can travel before they are stopped. The stopping power is about two orders of magnitude higher for carbon ions than it is for protons. Therefore, carbon needs considerably higher energies to reach the same depth. In the upper graph the range between 1 cm and 20 cm (i.e. the relevant one for radiotherapy) is indicated by the green bar at the right axis. In order to penetrate to this depth protons need energies between 32 MeV and 180 MeV, whereas carbon ions have to accelerated to energies of $60\ldots330$ MeV/u ($700\ldots4000$ MeV).

The second graph in Fig. 5.3 addresses the depth dose profile of protons and carbon ions. The energy per length that is deposited in water is shown as a function of penetration depth for a proton (p) and a carbon ion (C), with initial energies of 150 MeV and 250 MeV/u, respectively. Both species show the desired Bragg peak at the end of their path. The 1/e-width of the peak and the amount of energy inside this width is constant. A change in particle energy only affects the depth where this energy is deposited. For a single proton 830 keV lie within 17 μm, whereas a carbon ion loses 44 MeV within an 1/e-width of 81 μm. For an application in radiotherapy the *relative biological effectiveness* (RBE) has to be taken into account. The RBE is the dose-ratio of x-ray dose to particle dose to produce the same effect in biological tissue [145]. Protons have a RBE of about 1.1, which may be assumed to be almost independent on proton energy and penetration depth [146]. This changes significantly for carbon ions, the RBE increases from 1.6 in the entrance channel to about 5.9 at its Bragg peak [145]. This enhances the effect of carbon ions more strongly than for protons. The thin solid lines in the second graph of Fig. 5.3 show the dose for a carbon ion (CxRBE) and a proton (pxRBE) weighted by its respective RBE. If the average dose inside the 1/e-peaks are calculated to give the same biological effect, one arrives at the result, that about 54 protons (pxRBEx54) would give the same 1/e-dose average as one carbon ion. The carbon ions would therefore be more suitable to therapy, because of their improved ratio between high target dose and reduced integral dose.

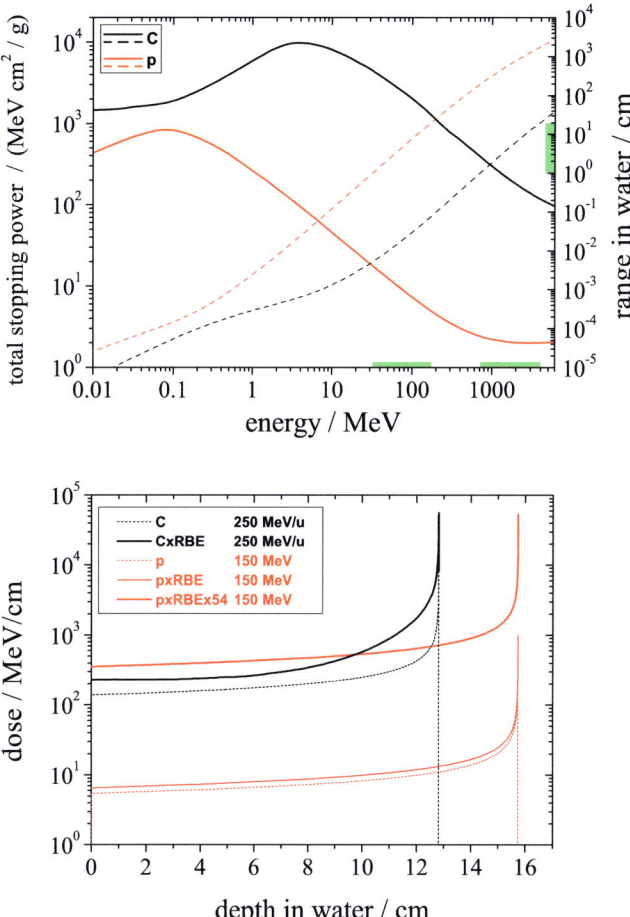

Figure 5.3: The stopping of protons and carbon ions in water. The upper graph shows energetic dependency of the total stopping power for protons and carbon ions in water as well as their range. The green bars show the energy intervals for protons and carbon ions corresponding to a penetration depth of 1 ... 20 cm. The lower graph shows the Bragg peaks representing the physical and biological dose for a proton and a carbon ion. For further explanations see text.

As of June 2004 almost 40000 patients suffering of inoperable tumors have been treated with proton beams and about 2000 patients received carbon beam treatment. The number of patients that can be treated is restricted by the limited quantity of proton therapy facilities, about 22 worldwide [147]. Protons are the "therapy of choice" for melanoma of the human eye and in paediatric oncology [148–151]. The main reason for the restriction to only a few therapy centers worldwide is the huge amount of investment necessary for such a facility consisting of the accelerator and the beam transport, that need to be radiation shielded. Large gantries are deployed in the beam transport to direct the therapeutic beam for tumor conformal irradiation. For carbon ions it is even more intricate, because of their higher magnetic rigidity and energies the accelerators for ions heavier than protons require more expansive accelerators and larger magnets in the beam transport lines [142]. Thus, there are presently no gantries available for carbon therapy.

With the ever increasing proton energies from laser plasma accelerators the feasibility of laser accelerated ions for radiotherapy in oncology was suggested [32, 152, 153]. With the experiment presented in Sec. 4.3 and 4.4 a proof of principle was given , that laser plasma accelerators are capable of producing monoenergetic proton beams, and therefore may be an option for the future of radiotherapy, given the progress in laser development and laser plasma science. Fig. 5.4 depicts this fact quite plainly, as the energy deposition in water for the peaked proton spectrum obtained from the irradiation of a dot (Fig. 4.6) and the simulated proton spectrum for the *Polaris*-scenario (Fig. 4.10). An extension of this concept to carbon ions seems to be conceivable, if corresponding targets with carbon dots can be produced and acceleration can be restricted to one species only by full disposal of the contamination layer. Using a different approach for target preparation, the laser acceleration of quasi-monoenergetic carbon ions has already been demonstrated [30].

As the acceleration of protons and ions in a laser plasma accelerator may be accomplished within only a few micrometers, the ions may be accelerated close to

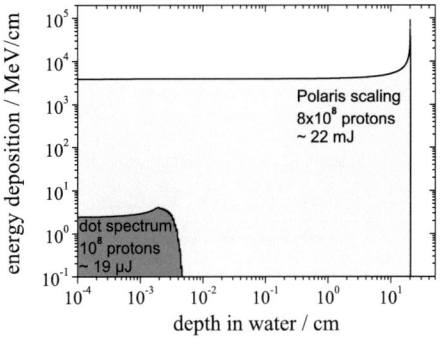
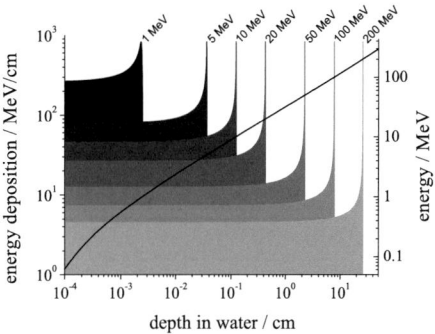

a) Energy deposition in water calculated from the experimental and simulated spectra in Ch. 4.

b) Energy deposition in water of one proton for different initial energies.

Figure 5.4: The energy deposition of the protons penetrating water. Please note the double logarithmic scale. (a) The protons from the peaked spectrum obtained from dot acceleration (Fig. 4.6) achieve a maximum range of only about 50 µm with a very low contrast peak at 20 µm, whereas the protons from the simulated spectrum for the *Polaris*-scenario (Fig. 4.10) can penetrate up to 20.4 cm into water with a high-contrast Bragg peak at 20.1 cm. (b) Increasing the proton energy leads to increased penetration depth, whereas the energy deposition in the 1/e-width of the Bragg peak is constant.

the target, reducing the effort of shielding to the target area only. Concepts for collimation and focusing of a laser generated proton beam are also at hand, using curved targets or laser-driven micro-lenses [5, 31]. Furthermore, since light has to be transported instead of particles, this approach would replace the magnetic beam transport and the heavy gantry by a construction of mirrors. Therefore, a facility providing therapeutic ion beams could, in principle, shrink to laboratory size and to affordable investments when using laser accelerators, and may therefore be more widely available than at a few sites worldwide, as it is currently the case with large accelerator facilities.

Bibliography

[1] V. Malka, S. Fritzler, E. Lefebvre, M. M. Aleonard, F. Burgy, J. P. Chambaret, J. F. Chemin, K. Krushelnick, G. Malka, S. P. D. Mangles, Z. Najmudin, M. Pittman, J. P. Rousseau, J. N. Scheurer, B. Walton, and A. E. Dangor. Electron acceleration by a wake field forced by an intense ultrashort laser pulse. *Science*, 298(5598):1596–1600, 2002.

[2] A. Modena, Z. Najmudin, A. E. Dangor, C. E. Clayton, K. A. Marsh, C. Joshi, V. Malka, C. B. Darrow, C. Danson, D. Neely, and F. N. Walsh. Electron acceleration from the breaking of relativistic plasma-waves. *Nature*, 377(6550):606–608, 1995.

[3] E. L. Clark, K. Krushelnick, J. R. Davies, M. Zepf, M. Tatarakis, F. N. Beg, A. Machacek, P. A. Norreys, M. I. K. Santala, I. Watts, and A. E. Dangor. Measurements of energetic proton transport through magnetized plasma from intense laser interactions with solids. *Physical Review Letters*, 84(4):670–673, 2000.

[4] A. Maksimchuk, S. Gu, K. Flippo, D. Umstadter, and V. Y. Bychenkov. Forward ion acceleration in thin films driven by a high-intensity laser. *Physical Review Letters*, 84(18):4108–4111, 2000.

[5] M. Roth, A. Blazevic, M. Geissel, T. Schlegel, T. E. Cowan, M. Allen, J. C. Gauthier, P. Audebert, J. Fuchs, J. Meyer-ter Vehn, M. Hegelich, S. Karsch, and A. Pukhov. Energetic ions generated by laser pulses: A detailed study on target properties. *Physical Review Special Topics - Accelerators and Beams*, 5(6):061301, 2002.

[6] R. A. Snavely, M. H. Key, S. P. Hatchett, T. E. Cowan, M. Roth, T. W. Phillips, M. A. Stoyer, E. A. Henry, T. C. Sangster, M. S. Singh, S. C.

Wilks, A. MacKinnon, A. Offenberger, D. M. Pennington, K. Yasuike, A. B. Langdon, B. F. Lasinski, J. Johnson, M. D. Perry, and E. M. Campbell. Intense high-energy proton beams from petawatt-laser irradiation of solids. *Physical Review Letters*, 85(14):2945–2948, 2000.

[7] F. Amiranoff. Fast electron production in ultra-short high-intensity laser-plasma interaction and its consequences. *Measurement Science & Technology*, 12(11):1795–1800, 2001.

[8] S. Düsterer, H. Schwoerer, W. Ziegler, C. Ziener, and R. Sauerbrey. Optimization of euv radiation yield from laser-produced plasma. *Applied Physics B – Lasers and Optics*, 73(7):693–698, 2001.

[9] H. Schwoerer, P. Gibbon, S. Düsterer, R. Behrens, C. Ziener, C. Reich, and R. Sauerbrey. MeV x rays and photoneutrons from femtosecond laser-produced plasmas. *Physical Review Letters*, 86(11):2317–2320, 2001.

[10] T. E. Cowan, A. W. Hunt, T. W. Phillips, S. C. Wilks, M. D. Perry, C. Brown, W. Fountain, S. Hatchett, J. Johnson, M. H. Key, T. Parnell, D. M. Pennington, R. A. Snavely, and Y. Takahashi. Photonuclear fission from high energy electrons from ultraintense laser-solid interactions. *Physical Review Letters*, 84(5):903–906, 2000.

[11] F. Ewald, H. Schwoerer, S. Düsterer, R. Sauerbrey, J. Magill, J. Galy, R. Schenkel, S. Karsch, D. Habs, and K. Witte. Application of relativistic laser plasmas for the study of nuclear reactions. *Plasma Physics and Controlled Fusion*, 45:A83–A91, 2003.

[12] K. W. D. Ledingham, P. McKenna, and R. P. Singhal. Applications for nuclear phenomena generated by ultra-intense lasers. *Science*, 300(5622): 1107–1111, 2003.

[13] H. Schwoerer, F. Ewald, R. Sauerbrey, J. Galy, J. Magill, V. Rondinella,

R. Schenkel, and T. Butz. Fission of actinides using a tabletop laser. *Europhysics Letters*, 61(1):47–52, 2003.

[14] K. Sokolowski-Tinten and D. von der Linde. Ultrafast phase transitions and lattice dynamics probed using laser-produced x-ray pulses. *Journal of Physics - Condensed Matter*, 16(49):R1517–R1536, 2004.

[15] S. Karsch, S. Düsterer, H. Schwoerer, F. Ewald, D. Habs, M. Hegelich, G. Pretzler, A. Pukhov, K. Witte, and R. Sauerbrey. High-intensity laser induced ion acceleration from heavy-water droplets. *Physical Review Letters*, 91(1):015001, 2003.

[16] G. Pretzler, A. Saemann, A. Pukhov, D. Rudolph, T. Schatz, U. Schramm, P. Thirolf, D. Habs, K. Eidmann, G. D. Tsakiris, J. Meyer-ter-Vehn, and K. J. Witte. Neutron production by 200 mJ ultrashort laser pulses. *Physical Review E*, 58(1):1165–1168, 1998.

[17] B. Liesfeld, K.-U. Amthor, F. Ewald, H. Schwoerer, J. Magill, J. Galy, G. Lander, and R. Sauerbrey. Nuclear reactions triggered by laser-accelerated relativistic electron jets. *Applied Physics B*, 79:1047–1052, 2004.

[18] B. Liesfeld. *A photon collider at relativistic intensities.* Dissertation, Friedrich-Schiller-Universität Jena, 2005.

[19] B. Liesfeld, J. Bernhardt, K.-U. Amthor, H. Schwoerer, and R. Sauerbrey. Single-shot autocorrelation at relativistic intensity. *Applied Physics Letters*, 86(1):161107, 2005.

[20] H. Schwoerer, B. Liesfeld, H.-P. Schlenvoigt, K.-U. Amthor, and R. Sauerbrey. Thomson-backscattered x rays from laser-accelerated electrons. *Physical Review Letters*, 96:014802, 2006.

[21] J. Faure, Y. Glinec, A. Pukhov, S. Kiselev, S. Gordienko, E. Lefebvre, J. P. Rousseau, F. Burgy, and V. Malka. A laser-plasma accelerator producing monoenergetic electron beams. *Nature*, 431(7008):541–544, 2004.

[22] C. G. R. Geddes, C. Toth, J. van Tilborg, E. Esarey, C. B. Schroeder, D. Bruhwilder, C. Nieter, J. Cary, and W. P. Leemans. High-quality electron beams from a laser wakefield accelerator using plasma-channel guiding. *Nature*, 431(7008):538–541, 2004.

[23] S. P. D. Mangles, C. D. Murphy, Z. Najmudin, A. G. R. Thomas, J. L. Collier, A. E. Dangor, E. J. Divall, P. S. Foster, J. G. Gallacher, C. J. Hooker, D. A. Jaroszynski, A. J. Langley, W. B. Mori, P. A. Norreys, F. S. Tsung, R. Viskup, B. R. Walton, and K. Krushelnick. Monoenergetic beams of relativistic electrons from intense laser-plasma interactions. *Nature*, 431 (7008):535–538, 2004.

[24] A. Pukhov and J. Meyer-ter Vehn. Laser wake field acceleration: the highly non-linear broken-wave regime. *Applied Physics B – Lasers and Optics*, 74 (4-5):355–361, 2002.

[25] A. Pukhov, S. Kiselev, I. Kostyukov, O. Shorokhov, and S. Gordienko. Relativistic laser-plasma bubbles: new sources of energetic particles and x-rays. *Nuclear Fusion*, 44(12):S191–S201, 2004.

[26] B. Hidding, K.-U. Amthor, B. Liesfeld, H. Schwoerer, S. Karsch, M. Geissler, L. Veisz, K. Schmid, J. G. Gallacher, S. P. Jamison, D. Jaroszynski, G. Pretzler, and Sauerbrey R. Generation of quasimonoenergetic electron bunches with 80-fs laser pulses. *Physical Review Letters*, 96:105004, 2006.

[27] K.-U. Amthor, B. Liesfeld, H. Schwoerer, R. Sauerbrey, and M. Geissler. Evolution of relativistic laser plasma channels – bubble-acceleration caught in the act. *to be published*, 2006.

[28] T. E. Cowan, J. Fuchs, H. Ruhl, A. Kemp, P. Audebert, M. Roth, R. Stephens, I. Barton, A. Blazevic, E. Brambrink, J. Cobble, J. Fernandez, J. C. Gauthier, M. Geissel, M. Hegelich, J. Kaae, S. Karsch, G. P. Le Sage, S. Letzring, M. Manclossi, S. Meyroneinc, A. Newkirk, H. Pepin,

and N. Renard-LeGalloudec. Ultralow emittance, multi-mev proton beams from a laser virtual-cathode plasma accelerator. *Physical Review Letters*, 92 (20):204801, 2004.

[29] H. Schwoerer, S. Pfotenhauer, O. Jäckel, K.-U. Amthor, B. Liesfeld, W. Ziegler, R. Sauerbrey, K. W. D. Ledingham, and T. Esirkepov. Laser-plasma acceleration of quasi-monoenergetic protons from micro-structured targets. *Nature*, 439:445–448, 2006.

[30] B. M. Hegelich, B. J. Albright, J. Cobble, K. Flippo, S. Letzring, M. Paffett, H. Ruhl, J. Schreiber, R. K. Schulze, and J. C. Fernandez. Laser acceleration of quasi-monoenergetic mev ion beams. *Nature*, 439(7075):441–444, 2006.

[31] T. Toncian, M. Borghesi, J. Fuchs, E. d'Humieres, P. Antici, P. Audebert, E. Brambrink, C. A. Cecchetti, A. Pipahl, L. Romagnani, and O. Willi. Ultrafast laser-driven microlens to focus and energy-select mega-electron volt protons. *Science*, 312(5772):410–413, April 2006.

[32] S. V. Bulanov and V. S. Khoroshkov. Feasibility of using laser ion accelerators in proton therapy. *Plasma Physics Reports*, 28(5):453–456, May 2002.

[33] T Tajima and G Mourou. Zettawatt-exawatt lasers and their applications in ultrastrong-field physics. *Physical Review Special Topics - Accelerators and Beams*, 5(3):031301, 2002.

[34] J. Bernhard. *Aufbau eines Experimentes zur Überlagerung zweier gegenläufiger intensiver Laserpulse*. Diplomarbeit, Friedrich-Schiller-Universität Jena, 2005.

[35] P. Gibbon. *Short Pulse Laser Interactions with Matter - An Introduction*. Imperial College Press London, 2005.

[36] E. S. Sarachik and G. T. Schappert. Classical theory of scattering of intense laser radiation by free electrons. *Physical Review D*, 1(10):2738–2753, 1970.

[37] J. D. Jackson. *Classical Electrodynamics*. Wiley, New York, 3rd edition, 1999.

[38] S. Y. Chen, A. Maksimchuk, and D. Umstadter. Experimental observation of relativistic nonlinear thomson scattering. *Nature*, 396(6712):653–655, 1998.

[39] E. Esarey, S. K. Ride, and P. Sprangle. Nonlinear thomson scattering of intense laser-pulses from beams and plasmas. *Physical Review E*, 48(4): 3003–3021, 1993.

[40] Y. Y. Lau, F. He, D. P. Umstadter, and R. Kowalczyk. Nonlinear thomson scattering: A tutorial. *Physics of Plasmas*, 10(5):2155–2162, 2003.

[41] D. Umstadter, S. Y. Chen, A. Maksimchuk, G. Mourou, and R. Wagner. Nonlinear optics in relativistic plasmas and laser wake field acceleration of electrons. *Science*, 273(5274):472–475, 1996.

[42] R. J. Goldston and P. H. Rutherford. *Introduction to Plasma Physics*. Institute of Physics Publishing, Bristol and Philadelphia, 1995.

[43] W. L. Kruer. *The physics of laser plasma interactions*. Frontiers in Physics. Westview Press, Boulder and Oxford, 2003.

[44] E. Esarey, J. Krall, and P. Sprangle. Envelope analysis of intense laser-pulse self-modulation in plasmas. *Physical Review Letters*, 72(18):2887–2890, 1994.

[45] C. E. Max, J. Arons, and A. B. Langdon. Self-modulation and self-focusing of electromagnetic waves in plasmas. *Physical Review Letters*, 33(4):209–212, 1974.

[46] G. Z. Sun, E. Ott, Y. C. Lee, and P. Guzdar. Self-focusing of short intense pulses in plasmas. *Physics of Fluids*, 30(2):526–532, 1987.

[47] E. Esarey, P. Sprangle, J. Krall, and A. Ting. Self-focusing and guiding of short laser pulses in ionizing gases and plasmas. *IEEE Journal of Quantum Electronics*, 33(11):1879–1914, 1997.

[48] R. Fedosejevs, X. F. Wang, and G. D. Tsakiris. Onset of relativistic self-focusing in high density gas jet targets. *Physical Review E*, 56(4):4615–4639, 1997.

[49] G. S. Sarkisov, V. Y. Bychenkov, V. N. Novikov, V. T. Tikhonchuk, A. Maksimchuk, S. Y. Chen, R. Wagner, G. Mourou, and D. Umstadter. Self-focusing, channel formation, and high-energy ion generation in interaction of an intense short laser pulse with a he jet. *Physical Review E*, 59(6): 7042–7054, 1999.

[50] B. Quesnel and P. Mora. Theory and simulation of the interaction of ultraintense laser pulses with electrons in vacuum. *Physical Review E*, 58(3): 3719–3732, 1998.

[51] T. Tajima and J. M. Dawson. Laser electron-accelerator. *Physical Review Letters*, 43(4):267–270, 1979.

[52] E. Esarey, P. Sprangle, J. Krall, and A. Ting. Overview of plasma-based accelerator concepts. *IEEE Transactions on Plasma Science*, 24(2):252–288, 1996.

[53] K. Krushelnick, Z. Najmudin, S. P. D. Mangles, A. G. R. Thomas, M. S. Wei, B. Walton, A. Gopal, E. L. Clark, A. E. Dangor, S. Fritzler, C. D. Murphy, P. A. Norreys, W. B. Mori, J. Gallacher, D. Jaroszynski, and R. Viskup. Laser plasma acceleration of electrons: Towards the production of monoenergetic beams. *Physics of Plasmas*, 12(5):056711, 2005.

[54] S. P. D. Mangles, B. R. Walton, M. Tzoufras, Z. Najmudin, R. J. Clarke, A. E. Dangor, R. G. Evans, S. Fritzler, A. Gopal, C. Hernandez-Gomez, W. B. Mori, W. Rozmus, M. Tatarakis, A. G. R. Thomas, F. S. Tsung, M. S. Wei, and K. Krushelnick. Electron acceleration in cavitated channels formed by a petawatt laser in low-density plasma. *Physical Review Letters*, 94(24):245001, 2005.

[55] P. Gibbon and E. Förster. Short-pulse laser-plasma interactions. *Plasma Physics and Controlled Fusion*, 38(6):769–793, 1996.

[56] C. Gahn, G. D. Tsakiris, A. Pukhov, J. Meyer-ter Vehn, G. Pretzler, P. Thirolf, D. Habs, and K. J. Witte. Multi-MeV electron beam generation by direct laser acceleration in high-density plasma channels. *Physical Review Letters*, 83(23):4772–4775, 1999.

[57] C. Gahn, G. D. Tsakiris, G. Pretzler, K. J. Witte, P. Thirolf, D. Habs, C. Delfin, and C. G. Wahlstrom. Generation of mev electrons and positrons with femtosecond pulses from a table-top laser system. *Physics of Plasmas*, 9(3):987–999, 2002.

[58] A. Pukhov, Z. M. Sheng, and J. Meyer-ter Vehn. Particle acceleration in relativistic laser channels. *Physics of Plasmas*, 6(7):2847–2854, 1999.

[59] S. Gordienko and A. Pukhov. Scalings for ultrarelativistic laser plasmas and quasimonoenergetic electrons. *Physics of Plasmas*, 12(4):043109, 2005.

[60] I. Kostyukov, A. Pukhov, and S. Kiselev. Phenomenological theory of laser-plasma interaction in "bubble" regime. *Physics of Plasmas*, 11(11):5256–5264, 2004.

[61] A. Pukhov, S. Gordienko, S. Kiselev, and I. Kostyukov. The bubble regime of laser-plasma acceleration: monoenergetic electrons and the scalability. *Plasma Physics and Controlled Fusion*, 46:B179–B186, 2004.

[62] S. P. Hatchett, C. G. Brown, T. E. Cowan, E. A. Henry, J. S. Johnson, M. H. Key, J. A. Koch, A. B. Langdon, B. F. Lasinski, R. W. Lee, A. J. Mackinnon, D. M. Pennington, M. D. Perry, T. W. Phillips, M. Roth, T. C. Sangster, M. S. Singh, R. A. Snavely, M. A. Stoyer, S. C. Wilks, and K. Yasuike. Electron, photon, and ion beams from the relativistic interaction of petawatt laser pulses with solid targets. *Physics of Plasmas*, 7(5):2076–2082, 2000.

[63] S. C. Wilks and W. L. Kruer. Absorption of ultrashort, ultra-intense laser light by solids and overdense plasmas. *IEEE Journal of Quantum Electronics*, 33(11):1954–1968, 1997.

[64] S. C. Wilks, W. L. Kruer, M. Tabak, and A. B. Langdon. Absorption of ultra-intense laser-pulses. *Physical Review Letters*, 69(9):1383–1386, 1992.

[65] K. Krushelnick, E. L. Clark, F. N. Beg, A. E. Dangor, Z. Najmudin, P. A. Norreys, M. Wei, and M. Zepf. High intensity laser-plasma sources of ions-physics and future applications. *Plasma Physics and Controlled Fusion*, 47: B451–B463, December 2005.

[66] S. C. Wilks, A. B. Langdon, T. E. Cowan, M. Roth, M. Singh, S. Hatchett, M. H. Key, D. Pennington, A. MacKinnon, and R. A. Snavely. Energetic proton generation in ultra-intense laser-solid interactions. *Physics of Plasmas*, 8(2):542–549, February 2001.

[67] S. J. Gitomer, R. D. Jones, F. Begay, A. W. Ehler, J. F. Kephart, and R. Kristal. Fast ions and hot-electrons in the laser-plasma interaction. *Physics of Fluids*, 29(8):2679–2688, 1986.

[68] M. Allen, Y. Sentoku, P. Audebert, A. Blazevic, T. Cowan, J. Fuchs, J. C. Gauthier, M. Geissel, M. Hegelich, S. Karsch, E. Morse, P. K. Patel, and M. Roth. Proton spectra from ultraintense laser-plasma interaction with thin foils: Experiments, theory, and simulation. *Physics of Plasmas*, 10(8): 3283–3289, 2003.

[69] M. Borghesi, D. H. Campbell, A. Schiavi, O. Willi, M. Galimberti, L. A. Gizzi, A. J. Mackinnon, R. D. Snavely, P. Patel, S. Hatchett, M. Key, and W. Nazarov. Propagation issues and energetic particle production in laser-plasma interactions at intensities exceeding 10^{19} W/cm^2. *Laser and Particle Beams*, 20(1):31–38, 2002.

[70] J. Fuchs, Y. Sentoku, S. Karsch, J. Cobble, P. Audebert, A. Kemp, A. Nikroo, P. Antici, E. Brambrink, A. Blazevic, E. M. Campbell, J. C. Fernandez, J. C. Gauthier, M. Geissel, M. Hegelich, H. Pepin, H. Popescu, N. Renard-LeGalloudec, M. Roth, J. Schreiber, R. Stephens, and T. E. Cowan. Comparison of laser ion acceleration from the front and rear surfaces of thin foils. *Physical Review Letters*, 94(4):045004, 2005.

[71] P. Mora. Plasma expansion into a vacuum. *Physical Review Letters*, 90(18): 185002, 2003.

[72] J. Fuchs, P. Antici, E. D'Humieres, E. Lefebvre, M. Borghesi, E. Brambrink, C. A. Cecchetti, M. Kaluza, V. Malka, M. Manclossi, S. Meyroneinc, P. Mora, J. Schreiber, T. Toncian, H. Pepin, and R. Audebert. Laser-driven proton scaling laws and new paths towards energy increase. *Nature Physics*, 2(1):48–54, 2006.

[73] M. Kaluza, J. Schreiber, M. I. K. Santala, G. D. Tsakiris, K. Eidmann, J. Meyer-ter Vehn, and K. J. Witte. Influence of the laser prepulse on proton acceleration in thin-foil experiments. *Physical Review Letters*, 93(4): 045003, 2004.

[74] A. P. L. Robinson, A. R. Bell, and R. J. Kingham. Effect of target composition on proton energy spectra in ultraintense laser-solid interactions. *Physical Review Letters*, 96(3):035005, 2006.

[75] M. H. Key, M. D. Cable, T. E. Cowan, K. G. Estabrook, B. A. Hammel, S. P. Hatchett, E. A. Henry, D. E. Hinkel, J. D. Kilkenny, J. A. Koch, W. L. Kruer, A. B. Langdon, B. F. Lasinski, R. W. Lee, B. J. MacGowan, A. MacKinnon, J. D. Moody, M. J. Moran, A. A. Offenberger, D. M. Pennington, M. D. Perry, T. J. Phillips, T. C. Sangster, M. S. Singh, M. A. Stoyer, M. Tabak, G. L. Tietbohl, M. Tsukamoto, K. Wharton, and S. C. Wilks. Hot electron production and heating by hot electrons in fast ignitor research. *Physics of Plasmas*, 5(5):1966–1972, 1998.

[76] E. d'Humieres, E. Lefebyre, L. Gremillet, and V. Malka. Proton acceleration mechanisms in high-intensity laser interaction with thin foils. *Physics of Plasmas*, 12(6):062704, 2005.

[77] T. Esirkepov, M. Yamagiwa, and T. Tajima. Laser ion-acceleration scaling laws seen in multi-parametric particle-in-cell simulations. *Physical Review Letters*, 96:105001, 2006.

[78] T. Z. Esirkepov, S. V. Bulanov, K. Nishihara, T. Tajima, F. Pegoraro, V. S. Khoroshkov, K. Mima, H. Daido, Y. Kato, Y. Kitagawa, K. Nagai, and S. Sakabe. Proposed double-layer target for the generation of high-quality laser-accelerated ion beams. *Physical Review Letters*, 89(17):175003, 2002.

[79] P. Gibbon, F. N. Beg, E. L. Clark, R. G. Evans, and M. Zepf. Tree-code simulations of proton acceleration from laser-irradiated wire targets. *Physics of Plasmas*, 11(8):4032–4040, 2004.

[80] J. J. Honrubia, M. Kaluza, J. Schreiber, G. D. Tsakiris, and J. Meyer-ter Vehn. Laser-driven fast-electron transport in preheated foil targets. *Physics of Plasmas*, 12(5):052708, 2005.

[81] A. Pukhov. Three-dimensional simulations of ion acceleration from a foil irradiated by a short-pulse laser. *Physical Review Letters*, 86(16):3562–3565, 2001.

[82] H. Ruhl, S. V. Bulanov, T. E. Cowan, T. V. Liseikina, P. Nickles, F. Pegoraro, M. Roth, and W. Sandner. Computer simulation of the three-dimensional regime of proton acceleration in the interaction of laser radiation with a thin spherical target. *Plasma Physics Reports*, 27(5):363–371, 2001.

[83] M. Borghesi, A. J. Mackinnon, D. H. Campbell, D. G. Hicks, S. Kar, P. K. Patel, D. Price, L. Romagnani, A. Schiavi, and O. Willi. Multi-MeV proton

source investigations in ultraintense laser-foil interactions. *Physical Review Letters*, 92(5):055003, 2004.

[84] M. Roth, E. Brambrink, P. Audebert, A. Blazevic, R. Clarke, J. Cobble, T.E. Cowan, J. Fernandez, J. Fuchs, M. Geissel, D. Habs, M. Hegelich, S. Karsch, K. Ledingham, D. Neely, H. Ruhl, T. Schlegel, and J. Schreiber. Laser accelerated ions and electron transport in ultra-intense laser matter interaction. *Laser and Particle Beams*, 23:95–100, 2005.

[85] J. Schreiber, M. Kaluza, F. Grüner, U. Schramm, B. M. Hegelich, J. Cobble, M. Geissler, E. Brambrink, J. Fuchs, P. Audebert, D. Habs, and K. Witte. Source-size measurements and charge distributions of ions accelerated from thin foils irradiated by high-intensity laser pulses. *Applied Physics B – Lasers and Optics*, 79(8):1041–1045, 2004.

[86] V. Malka, J. Faure, Y. Glinec, A. Pukhov, and J. P. Rousseau. Monoenergetic electron beam optimization in the bubble regime. *Physics of Plasmas*, 12(5):056702, 2005.

[87] E. Miura, K. Koyama, S. Kato, N. Saito, M. Adachi, Y. Kawada, T. Nakamura, and M. Tanimoto. Demonstration of quasi-monoenergetic electron-beam generation in laser-driven plasma acceleration. *Applied Physics Letters*, 86(25):251501, 2005.

[88] A. Yamazaki, H. Kotaki, I. Daito, M. Kando, S. V. Bulanov, T. Zh. Esirkepov, S. Kondo, S. Kanazawa, T. Homma, K. Nakajima, Y. Oishi, T. Nayuki, T. Fujii, and K. Nemoto. Quasi-monoenergetic electron beam generation during laser pulse interaction with very low density plasmas. *Physics of Plasmas*, 12(9):093101, 2005.

[89] F. Ewald. *Harte Röntgenstrahlung aus relativistischen Laserplasmen und laserinduzierte Kernreaktionen*. Dissertation, Friedrich-Schiller-Universität Jena, 2004.

[90] C. Ziener. *Aufbau eines 12 Terawatt Titan:Saphir-Lasers zur effizienten Erzeugung charakteristischer Röntgenstrahlung.* Dissertation, Friedrich-Schiller-Universität Jena, 2001.

[91] P. Maine, D. Strickland, P. Bado, M. Pessot, and G. Mourou. Generation of ultrahigh peak power pulses by chirped pulse amplification. *IEEE Journal of Quantum Electronics*, 24(2):398–403, 1988.

[92] O. E. Martinez. Design of high-power ultrashort pulse-amplifiers by expansion and recompression. *IEEE Journal of Quantum Electronics*, 23(8): 1385–1387, 1987.

[93] D. Strickland and G. Mourou. Compression of amplified chirped optical pulses. *Optics Communications*, 55(6):447–449, 1985.

[94] D. Albach. *Aufbau einer Apparatur zur Vorpulsunterdrückung eines 15 Terawatt-Titan:Saphir-Lasers.* Diplomarbeit, Friedrich-Schiller-Universität Jena, 2005.

[95] R. Benattar, C. Popovics, and R. Sigel. Polarized-light interferometer for laser fusion studies. *Review of Scientific Instruments*, 50(12):1583–1585, 1979.

[96] I. H. Hutchinson. *Principles of plasma diagnostics.* Cambridge University Press, 1987.

[97] P. Monot, T. Auguste, P. Gibbon, F. Jakober, G. Mainfray, A. Dulieu, M. Louisjacquet, G. Malka, and J. L. Miquel. Experimental demonstration of relativistic self-channeling of a multiterawatt laser-pulse in an underdense plasma. *Physical Review Letters*, 74(15):2953–2956, 1995.

[98] M. Geissler. private communication, 2005.

[99] M. Geissler. *Interaction of ultrashort high-power laser pulses with plasmas.* Dissertation, TU Wien, 2000.

[100] M. J. Berger, J. S. Coursey, M. A. Zucker, and J. Chang. Stopping-power and range tables for electrons, protons, and helium ions. *NIST Standard Reference Database 124*, http://physics.nist.gov/PhysRefData/contents-radi.html, 2005.

[101] B. Hidding. private communication, 2005.

[102] K. A. Tanaka, T. Yabuuchi, T. Sato, R. Kodama, Y. Kitagawa, T. Takahashi, T. Ikeda, Y. Honda, and S. Okuda. Calibration of imaging plate for high energy electron spectrometer. *Review of Scientific Instruments*, 76(1): 013507, 2005.

[103] K.-U. Amthor. *Plasmadiagnose in Experimenten zur Wechselwirkung intensiver Laserimpulse mit Materie*. Diplomarbeit, Friedrich-Schiller-Universität Jena, 2002.

[104] M. Born and E. Wolf. *Principles of optics*. Cambridge University Press, 1999.

[105] A. J. Mackinnon, M. Borghesi, R. Gaillard, G. Malka, O. Willi, A. A. Offenberger, A. Pukhov, J. Meyer-ter Vehn, B. Canaud, J. L. Miquel, and N. Blanchot. Intense laser pulse propagation and channel formation through plasmas relevant for the fast ignitor scheme. *Physics of Plasmas*, 6(5):2185–2190, 1999.

[106] S. Fritzler, V. Malka, G. Grillon, J. P. Rousseau, F. Burgy, E. Lefebvre, E. d'Humieres, P. McKenna, and K. W. D. Ledingham. Proton beams generated with high-intensity lasers: Applications to medical isotope production. *Applied Physics Letters*, 83(15):3039–3041, 2003.

[107] O. Jäckel. *Vermessung von Ionenspektren aus relativistischen laserproduzierten Plasmen*. Diplomarbeit, Friedrich-Schiller-Universität Jena, 2006.

[108] A. Waheed, S. Manzoor, R. Cherubini, G. Moschini, L. Lembo, and H. A. Khan. A more suitable etching condition to register low energy proton tracks

in CR39 for neutron dosimetry. *Nuclear Instruments and Methods in Physics Research B*, 47:320–328, 1990.

[109] J. Badziak, E. Woryna, R. Parys, K. Y. Platonov, S. Jablonski, L. Rye, A. B. Vankov, and J. Wolowski. Fast proton generation from ultrashort laser pulse interaction with double-layer foil targets. *Physical Review Letters*, 8721(21): 215001, 2001.

[110] H. Kishimura, H. Morishita, Y. H. Okano, Y. Okano, Y. Hironaka, K. Kondo, K. G. Nakamura, Y. Oishi, and K. Nemoto. Enhanced generation of fast protons from a polymer-coated metal foil by a femtosecond intense laser field. *Applied Physics Letters*, 85(14):2736–2738, 2004.

[111] S. Pfotenhauer. *Generation of Quasi-Monoenergetic Proton Beams from High Intensity Laser Plasmas*. Diplomarbeit, Friedrich-Schiller-Universität Jena, 2006.

[112] M. Hegelich, S. Karsch, G. Pretzler, D. Habs, K. Witte, W. Guenther, M. Allen, A. Blazevic, J. Fuchs, J. C. Gauthier, M. Geissel, P. Audebert, T. Cowan, and M. Roth. Mev ion jets from short-pulse-laser interaction with thin foils. *Physical Review Letters*, 89(8):085002, 2002.

[113] A. J. Mackinnon, M. Borghesi, S. Hatchett, M. H. Key, P. K. Patel, H. Campbell, A. Schiavi, R. Snavely, S. C. Wilks, and O. Willi. Effect of plasma scale length on multi-mev proton production by intense laser pulses. *Physical Review Letters*, 86(9):1769–1772, 2001.

[114] T. Esirkepov. private communication, 2005.

[115] K. Matsukado, T. Esirkepov, K. Kinoshita, H. Daido, T. Utsumi, Z. Li, A. Fukumi, Y. Hayashi, S. Orimo, M. Nishiuchi, S. V. Bulanov, T. Tajima, A. Noda, Y. Iwashita, T. Shirai, T. Takeuchi, S. Nakamura, A. Yamazaki, M. Ikegami, T. Mihara, A. Morita, M. Uesaka, K. Yoshii, T. Watanabe, T. Hosokai, A. Zhidkov, A. Ogata, Y. Wada, and T. Kubota. Energetic

protons from a few-micron metallic foil evaporated by an intense laser pulse. *Physical Review Letters*, 91(21):215001, November 2003.

[116] J. Hein, S. Podleska, M. Siebold, M. Hellwing, R. Bodefeld, R. Sauerbrey, D. Ehrt, and W. Wintzer. Diode-pumped chirped pulse amplification to the joule level. *Applied Physics B – Lasers and Optics*, 79(4):419–422, September 2004.

[117] S. Fritzler, K. Ta Phuoc, V. Malka, A. Rousse, and E. Lefebvre. Ultrashort electron bunches generated with high-intensity lasers: Applications to injectors and x-ray sources. *Applied Physics Letters*, 83(19):3888–3890, 2003.

[118] K. Krushelnick, E. L. Clark, R. Allott, F. N. Beg, C. N. Danson, A. Machacek, V. Malka, Z. Najmudin, D. Neely, P. A. Norreys, M. R. Salvati, M. I. K. Santala, M. Tatarakis, I. Watts, M. Zepf, and A. E. Dangor. Ultrahigh-intensity laser-produced plasmas as a compact heavy ion injection source. *IEEE Transactions on Plasma Science*, 28(4):1184–1189, 2000.

[119] Stanley Humphries. *Principles of Charged Particle Acceleration*. John Wiley and Sons, 1986.

[120] W. P. Leemans, C. G. R. Geddes, J. Faure, C. Toth, J. van Tilborg, C. B. Schroeder, E. Esarey, G. Fubiani, D. Auerbach, B. Marcelis, M. A. Carnahan, R. A. Kaindl, J. Byrd, and M. C. Martin. Observation of terahertz emission from a laser-plasma accelerated electron bunch crossing a plasma-vacuum boundary. *Physical Review Letters*, 91(7):074802, 2003.

[121] W. P. Leemans, E. Esarey, J. van Tilborg, P. A. Michel, C. B. Schroeder, C. Toth, C. G. R. Geddes, and B. A. Shadwick. Radiation from laser accelerated electron bunches: Coherent terahertz and femtosecond x-rays. *IEEE Transactions on Plasma Science*, 33(1):8–22, 2005.

[122] J. van Tilborg, C. B. Schroeder, C. V. Filip, C. Toth, C. G. R. Geddes, G. Fubiani, R. Huber, R. A. Kaindl, E. Esarey, and W. P. Leemans. Temporal

characterization of femtosecond laser-plasma-accelerated electron bunches using terahertz radiation. *Physical Review Letters*, 96(1):014801, 2006.

[123] S. Karsch, K.-U. Amthor, A. Debus, B. Liesfeld, H. Schwoerer, R. Sauerbrey, S. Jamison, J. Gallacher, C. J. Murphy, and D. Jaroczynski. Bunch length measurements of laser accelerated electron using thz transition radiation. *to be published.*

[124] D. Kaganovich, A. Ting, D. F. Gordon, R. F. Hubbard, T. G. Jones, A. Zigler, and P. Sprangle. First demonstration of a staged all-optical laser wakefield acceleration. *Physics of Plasmas*, 12(10):100702, 2005.

[125] W. P. Leemans, J. van Tilborg, J. Faure, C. G. R. Geddes, C. Toth, C. B. Schroeder, E. Esarey, G. Fubiani, and G. Dugan. Terahertz radiation from laser accelerated electron bunches. *Physics of Plasmas*, 11(5):2899–2906, 2004.

[126] 4GLS conceptual design report – Light years ahead. http://www.4gls.ac.uk, 2006.

[127] K. Boyer, T. S. Luk, and C. K. Rhodes. Possibility of optically induced nuclear-fission. *Physical Review Letters*, 60(7):557–560, 1988.

[128] K. W. D. Ledingham, I. Spencer, T. McCanny, R. P. Singhal, M. I. K. Santala, E. Clark, I. Watts, F. N. Beg, M. Zepf, K. Krushelnick, M. Tatarakis, A. E. Dangor, P. A. Norreys, R. Allott, D. Neely, R. J. Clark, A. C. Machacek, J. S. Wark, A. J. Cresswell, D. C. W. Sanderson, and J. Magill. Photonuclear physics when a multiterawatt laser pulse interacts with solid targets. *Physical Review Letters*, 84(5):899–902, 2000.

[129] K. W. D. Ledingham, J. Magill, P. McKenna, J. Yang, J. Galy, R. Schenkel, J. Rebizant, T. McCanny, S. Shimizu, L. Robson, R. P. Singhal, M. S. Wei, S. P. D. Mangles, P. Nilson, K. Krushelnick, R. J. Clarke, and P. A. Nor-

reys. Laser-driven photo-transmutation of I-129 - a long-lived nuclear waste product. *Journal of Physics D - Applied Physics*, 36(18):L79–L82, 2003.

[130] W. P. Leemans, D. Rodgers, P. E. Catravas, C. G. R. Geddes, G. Fubiani, E. Esarey, B. A. Shadwick, R. Donahue, and A. Smith. Gamma-neutron activation experiments using laser wakefield accelerators. *Physics of Plasmas*, 8(5):2510–2516, 2001.

[131] G. Malka, M. M. Aleonard, J. F. Chemin, G. Claverie, M. R. Harston, J. N. Scheurer, V. Tikhonchuk, S. Fritzler, V. Malka, P. Balcou, G. Grillon, S. Moustaizis, L. Notebaert, E. Lefebvre, and N. Cochet. Relativistic electron generation in interactions of a 30 TW laser pulse with a thin foil target. *Physical Review E*, 66(6):066402, 2002.

[132] R. Behrens, H. Schwoerer, S. Düsterer, P. Ambrosi, G. Pretzler, S. Karsch, and R. Sauerbrey. A thermoluminescence detector-based few-channel spectrometer for simultaneous detection of electrons and photons from relativistic laser-produced plasmas. *Review of Scientific Instruments*, 74(2):961–968, 2003.

[133] T. W. Phillips, M. D. Cable, T. E. Cowan, S. P. Hatchett, E. A. Henry, M. H. Key, M. D. Perry, T. C. Sangster, and M. A. Stoyer. Diagnosing hot electron production by short pulse, high intensity lasers using photonuclear reactions. *Review of Scientific Instruments*, 70(1):1213–1216, 1999.

[134] I. Spencer, K. W. D. Ledingham, R. P. Singhal, T. McCanny, P. McKenna, E. L. Clark, K. Krushelnick, M. Zepf, F. N. Beg, M. Tatarakis, A. E. Dangor, R. D. Edwards, M. A. Sinclair, P. A. Norreys, R. J. Clarke, and R. M. Allott. A nearly real-time high temperature laser-plasma diagnostic using photonuclear reactions in tantalum. *Review of Scientific Instruments*, 73 (11):3801–3805, 2002.

[135] Handbook on photonuclear data for applications – cross sections and spectra. Technical report, IAEA, 2000.

[136] G. H. McCall. Calculation of x-ray bremsstrahlung and characteristic line emission produced by a Maxwellian electron-distribution. *Journal of Physics D - Applied Physics*, 15(5):823–831, 1982.

[137] J. Magill, H. Schwoerer, F. Ewald, J. Galy, R. Schenkel, and R. Sauerbrey. Laser transmutation of iodine-129. *Applied Physics B – Lasers and Optics*, 77(4):387–390, 2003.

[138] M. Borghesi, D. H. Campbell, A. Schiavi, O. Willi, A. J. Mackinnon, D. Hicks, P. Patel, L. A. Gizzi, M. Galimberti, and R. J. Clarke. Laser-produced protons and their application as a particle probe. *Laser and Particle Beams*, 20(2):269–275, 2002.

[139] M. Borghesi, P. Audebert, S. V. Bulanov, T. Cowan, J. Fuchs, J. C. Gauthier, A. J. MacKinnon, P. K. Patel, G. Pretzler, L. Romagnani, A. Schiavi, T. Toncian, and O. Willi. High-intensity laser-plasma interaction studies employing laser-driven proton probes. *Laser and Particle Beams*, 23(3): 291–295, 2005.

[140] M. Roth, T. E. Cowan, M. H. Key, S. P. Hatchett, C. Brown, W. Fountain, J. Johnson, D. M. Pennington, R. A. Snavely, S. C. Wilks, K. Yasuike, H. Ruhl, F. Pegoraro, S. V. Bulanov, E. M. Campbell, M. D. Perry, and H. Powell. Fast ignition by intense laser-accelerated proton beams. *Physical Review Letters*, 86(3):436–439, 2001.

[141] R. R. Wilson. Radiological use of fast protons. *Radiology*, 47(5):487–491, 1946.

[142] G. Kraft. Tumor therapy with heavy charged particles. *Progress in Particle and Nuclear Physics*, 45:S473–S544, 2000.

[143] E. B. Hug. Protons versus photons: A status assessment at the beginning of the 21(st) century. *Radiotherapy and Oncology*, 73:S35–S37, 2004.

[144] J. F. Ziegler and J. P. Biersack. Stopping and range of ions in matter – SRIM. http://www.srim.org/.

[145] W. K. Weyrather and G. Kraft. RBE of carbon ions: Experimental data and the strategy of RBE calculation for treatment planning. *Radiotherapy and Oncology*, 73:S161–S169, 2004.

[146] P. Scampoli. Biological effects of accelerated protons. *Radiotherapy and Oncology*, 73:S130–S133, 2004.

[147] J. Sisterson. Ion beam therapy in 2004. *Nuclear Instruments & Methods in Physics Research Section B – Beam Interactions with Materials and Atoms*, 241(1-4):713–716, December 2005.

[148] B. Damato, A. Kacperek, M. Chopra, Campbell I. R., and R. D. Errington. Proton beam radiotherapy of choroidal melanoma: The Liverpool–Clatterbridge experience. *International Journal of Radiation Oncology Biology Physics*, 62:1405–1411, 2005.

[149] M. J. McDerby, N. W. John, J. N. H. Brunt, and A. Kacperek. Modelling concepts of proton eye radiotherapy. *Physiological Measurement*, 22:611–623, 2001.

[150] F. Saran. New technology for radiotherapy in paediatric oncology. *European Journal of Cancer*, 40(14):2091–2105, 2004.

[151] V. C. Wilson, J. McDonough, and Z. Tochner. Proton beam irradiation in pediatric oncology – an overview. *Journal of Pediatric Hematology Oncology*, 27(8):444–448, 2005.

[152] S. V. Bulanov, T. Z. Esirkepov, V. S. Khoroshkov, A. V. Kunetsov, and F. Pegoraro. Oncological hadrontherapy with laser ion accelerators. *Physics Letters A*, 299(2-3):240–247, 2002.

[153] V. Malka, S. Fritzler, E. Lefebvre, E. d'Humieres, R. Ferrand, G. Grillon, C. Albaret, S. Meyroneinc, J. P. Chambaret, A. Antonetti, and D. Hulin. Practicability of protontherapy using compact laser systems. *Medical Physics*, 31(6):1587–1592, June 2004.

Acknowledgements

The lion's share of my gratitude has been earned by the people constituting the *Institute of Optics and Quantum Electronics* at the Friedrich-Schiller-University in Jena, where I learned and studied the trade of laser plasma physics.

I would like to thank Prof. Dr. Roland Sauerbrey for introducing me to this fascinating domain of physics and for excellently supervising my thesis work. I am most grateful to Dr. Heinrich Schwoerer, who has been my mentor and guide during my years at the IOQ. I am still amazed by the effectiveness of Dr. Ben Liesfeld, with whom I had the pleasure to share office and lab, experiment on laser plasma, and discuss physics as well as what can be done with the help of IT and any gadget that we came across. I highly appreciate the acuteness, enthusiasm and assistance of Oliver Jäckel and Sebastian Pfotenhauer, who will continue to pursue laser proton acceleration. Nonetheless, I thank Falk Ronneberger and Burgarg Beleites for their tender loving care towards JETI to keep everything running and ready, and Wolfgang Ziegler the technology wizard of IOQ, who attended to anything from mechanics to targets, which he produced together with Jens Polz. I am also obliged to the staff of the electronic and mechanic workshops.

Furthermore, I would like to thank the Transregional Collaborative Research Centre of the German Research Foundation *Relativistic Laser Plasma Dynamics*, Transregio 18, which is a driving force in the field of laser plasma interaction in Germany. Amongst those how I had the opportunity of cooperating with, I would like to thank Prof. Dr. Georg Pretzler, Dr. Stefan Karsch and Bernhard Hidding. Thanks are especially due to Dr. Michael Geissler for *illuminating* the processes of laser plasma acceleration. I am indebted to Prof. Ken Ledingham and Prof. Timur Esirkepov for their conception of monoenergetic protons. Moreover I am thankful to Dr. Friederike Ewald and Jörg Schreiber for their perspective on things.

Last but most important, I would like to thank my friends, my family and Mareike for their constant love and support.

Zusammenfassung

Um geladene Teilchen zu beschleunigen benötigt man elektrische Felder. Die maximalen elektrischen Felder in Hochfrequenz-Teilchenbeschleunigern liegen bei mehreren $10\,\mathrm{MV/m}$. Dieses Feld ist wegen der Zerstörung der Elektroden durch einsetzende Ionisation nach oben begrenzt, d. h. das Material geht in den Plasmazustand über. Was wäre aber, wenn man diese Limitierung umgehen könnte, indem man das Beschleunigungsfeld von vorneherein in einem Plasma erzeugt?

Der Vorschlag im elektrischen Feld einer laser-induzierten Plasmawelle Elektronen zu beschleunigen wurde schon vor einem guten viertel Jahrhundert gemacht. In einem Plasma liegt das maximal erreichbare elektrische Feld im Bereich von $10\,\mathrm{GV/cm}$, d. h. um fünf Größenordnungen höher als in einem Hochfrequenz-Teilchenbeschleuniger. Die Strecken, die notwendig sind, um hochenergetische Teilchen zu erzeugen, schrumpfen daher auf die Größe von Millimetern. Laser-Plasma-Beschleunigung könnte daher zur Technologie der zukünftigen Teilchenbeschleunigung werden.

Allerdings sind die Prozesse in einem Laser-Plasma-Beschleuniger (LPB) geprägt von Nicht-Linearitäten und Instabilitäten, was die Kontrollierbarkeit der so erzeugten Teilchenstrahlen beeinträchtigt. In den letzten Jahren gelang es weltweit mehrfach, die Idee der laser-induzierten Teilchenbeschleunigung auch experimentell umzusetzen. Elektronen konnten so bis auf Energien von mehreren hundert MeV beschleunigt werden. Weit anspruchsvoller als immer höhere Energien zu erreichen, ist allerdings die Erzeugung von monoenergetischen Teilchenstrahlen aus dem Plasma. Für Elektronen konnte vor zwei Jahren von mehreren Gruppen gezeigt werden, daß auch die Anforderung der Schmalbandigkeit erfüllt werden kann. Laser-induzierte Teilchenstrahlen haben zusätzlich noch die Eigenschaft, daß sie stark gebündelt sind. Hierfür waren Laserimpulse kürzer als 50 fs und dünne Plasmen ($n_\mathrm{e} \sim 10^{19}\,\mathrm{cm}^{-3}$) notwendig. Schmalbandigkeit und gute Kollimation erhöhen die potentiellen Anwendungsmöglichkeiten der LPB um ein Vielfaches.

In der vorliegenden Arbeit werden LPB, die am Jenaer Titan:Saphir Laser (JETI) untersucht worden, vorgestellt. Der LPB für Elektronen basiert auf der Wechselwirkung hoch-intensiver Laserimpulse mit einem gepulsten Helium Gasstrahl. Der Laserimpuls wird in dieses Gas fokussiert, wo er aufgrund seiner hohen Intensität die Heliumatome vollständig ionisiert. Die nichtlineare Wechselwirkung zwischen dem Laserimpuls und dem Plasma führt zu Selbst-Fokussierung und der Ausbildung eines relativistischen Kanals. Im Rahmen dieser Arbeit wurde dieser LPB charakterisiert und es konnte gezeigt werden, daß auch für ein erweitertes Parameterregime durch selbstmodulierte Laserimpulse quasi-monoenergetische Elektronenstrahlen erzeugt werden können.

Die Charakterisierung des Elektronenbeschleunigers beinhaltete eine optische Diagnose der Entwicklung der Laser-Plasma-Wechselwirkung mit hoher zeitlicher Auflösung. In Anrege-Abfrage-Experimenten wurden Schattenbilder des relativistischen Kanals aufgenommen. Diese Diagnostik ermöglichte die erste Beobachtung des Elektronenbündels, das in einer Bubble beschleunigt wird. Das Elektronenbündel beeinflußt die Ausbreitung des Abfrageimpulses derart, daß in den Schattenbildern eine Struktur aus Interferenzringen beobachtet werden konnte, die auf die Position des Elektronenbündels in der Bubble schließen läßt.

Der LPB für Protonen sieht etwas anders aus. Protonen sind träger als Elektronen, deswegen müssen die beschleunigenden Felder länger aufrecht erhalten werden. Hierfür verwendet man dünne Metallfolien, die man mit den hoch-intensiven Laserpulsen bestrahlt. Es entsteht wiederum ein Plasma, in welchem Elektronen beschleunigt werden. Dies geschieht vorzugsweise in Richtung der Laserausbreitung bzw. in Richtung der Metallfolie. Die Folien werden so dünn gewählt, daß sie die Elektronen nahezu ungestört durchdringen können. An der Rückseite der Folie werden durch Ionisation Protonen und Ionen erzeugt und zwischen der Elektronenwolke und der Folienrückseite bildet sich ein quasi-elektrostatisches Feld aus. In diesem Feld werden vornehmlich Protonen beschleunigt, da diese das größte Ladungs-Masse-Verhältnis haben.

Ebenfalls im Rahmen dieser Arbeit wurden die für Protonenbeschleunigung optimalen Parameter für den JETI und die verwendeten Metallfolien identifiziert. Dafür war es förderlich, eine Detektionsmethode für Protonenspektren zu implementieren, die der hohen Wiederholrate des Lasersystems (10 Hz) gerecht wird. Der Einsatz einer bildgebenden Mikrokanalplatte, die mittels eines PCs über eine Kamera in Echtzeit ausgelesen werden kann, ermöglichte die systematische, schnelle und flexible Untersuchung der Protonbeschleunigung für verschiedene Laser- und Metallfolien-Parameter. Einem theoretischen Vorschlag folgend wurde so die Protonenbeschleunigung von speziell beschichteten Metallfolien untersucht. Das Beschleunigungsfeld an der Folienrückseite ist, im Bereich des Laserfokus, nahezu homogen und fällt außerhalb dieses Bereichs stark ab. Theoretische Betrachtungen und Simulationen sagten voraus, daß Folien, die auf der Rückseite einen Punkt tragen, der reich an Protonen ist und nur von der Größe des homogenen Bereichs ist, zur Erzeugung monoenergetischer Protonen verwendet werden können. In einem Experiment in dieser Arbeit wird dieses Prinzip erstmal experimentell demonstriert. Kleine Polymer-Punkte auf der Folienrückseite erhöhen die Protonenzahl im Zentrum des Beschleunigungsfeldes und führten so zu Protonenspektren, die schmalbandige Strukturen auswiesen.

Mit den Untersuchungen zu den LPB für Elektronen und Protonen konnte ein wichtiger Beitrag zum Verständnis der laser-induzierten Teilchenbeschleunigung gewonnen werden. Die Demonstration, daß LPB prinzipiell in der Lage sind kollimierte, hoch-energetische, monochromatische Teilchenstrahlen zu erzeugen, zeichnet den Weg zur zukünftigen Anwendbarkeit der LPB als Injektor für herkömmliche Beschleuniger, als erschwingliche Teilchenbeschleuniger in der universitären Forschung oder in der Medizin.